PRACTICAL MATH

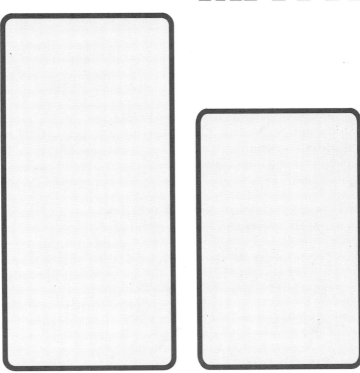

AMERICAN TECHNICAL PUBLISHERS, INC.
HOMEWOOD, ILLINOIS 60430

Staff

1 2 3 4 5 6 7 8 9 – 94 – 9 8 7 6 5 4 3 2 1

Printed in the United States of America

ISBN 0-8269-2244-9

INTRODUCTION

Practical Math is a basic mathematics book with essential text, many illustrations, Examples, Practice Problems, Reviews, and Tests. The questions are designed to represent typical problems that must be solved by tradesworkers everyday on the job. A comprehensive chart is provided on page 302 to explain the mathematical symbols used throughout this book.

Each of the twelve chapters contains an Introduction providing an overview of chapter content. The chapter content includes Examples of specific mathematical processes and Practice Problems for the student to answer. The student should use a separate sheet of paper to answer all Practice Problems. The answers to all odd-numbered Practice Problems are given in the back of this book.

Each chapter concludes with a Review to test comprehension and a Test to verify content application. Space is provided in the Review and Test to record all answers. Unless otherwise indicated, always round all decimal answers to two places (hundredths).

The Publishers

CONTENTS

WHOLE NUMBERS

W hole numbers are numbers that have no fractional or decimal parts. Arabic and Roman numerals are the two most common systems used for calculations and notations. Processes used for the calculation of whole numbers include adding, subtracting, multiplying, and dividing.

WHOLE NUMBERS

Whole numbers, or *integers*, are numbers that have no fractional or decimal parts. For example, numbers such as 1, 2, 5, 14, 35, 71, 144, 1966, etc. are whole numbers. They are the numbers used for counting all things that can be counted as separate objects or entities, such as chairs, trees, automobiles, bolts, doors, pipes, etc.

There are many more numbers that are partially fractional or decimal than there are integers, because most quantities cannot be directly counted, but must be measured in some way. *Continuous quantities* are the numbers resulting from measurements that cannot be directly counted.

Whole numbers are either odd or even numbers. See Figure 1-1. An *odd number* is any number that cannot be divided by 2 an exact number of times. An odd number always ends in 1, 3, 5, 7, or 9. For example, numbers such as 3, 35, 155, 2901, etc. are odd numbers.

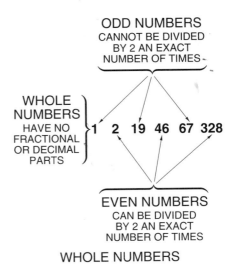

Figure 1-1. Whole numbers have no fractional or decimal parts.

1

An *even number* is any number that can be divided by 2 an exact number of times. An even number is any number ending in 2, 4, 6, 8, or 0. For example, the numbers 8, 32, 144, 1996, etc. are even numbers.

Prime numbers are numbers that can be divided an exact number of times, only by themselves and the number 1. For example, the numbers 1, 2, 3, 5, 7, 11, 13, 19, 23, etc. are prime numbers.

Arabic and Roman numerals are the two most common systems used for calculations and notations. Arabic numerals use ten digits, and Roman numerals use letters to represent numerical values.

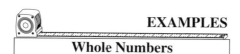

EXAMPLES
Whole Numbers

1. List five whole numbers.

A whole number is any number that has no fractional or decimal parts. For example, the numbers 8, 15, 467, 31, and 88 are typical whole numbers.

2. List five integers.

By definition, an integer is the same as a whole number. Any whole number is an integer. For ex-ample, the numbers 8, 15, 467, 31, and 88 are typical integers.

3. List five odd numbers.

An odd number is one that can-not be divided by 2 without leaving a remainder. For example, the numbers 31, 57, 11, 13, and 61 are all odd numbers.

4. List five even numbers.

An even number can be divided by 2 without leaving a remainder. For example, the numbers 2, 10, 50, 34, and 58 are even numbers.

5. List five prime numbers.

A prime number is a number that can only be divided an exact number of times by itself and 1. For example, the numbers 53, 59, 61, and 67 are all prime numbers.

PRACTICE PROBLEMS
Whole Numbers

1. List the whole numbers be-tween 23 and 37.
2. List the integers from 1 to 12.
3. List the odd numbers from 17 to 41.
4. List the even numbers from 40 to 60.
5. List the prime numbers from 13 to 43.

Arabic Numerals

Arabic numerals are expressed by the ten digits 0, 1, 2, 3, 4, 5, 6, 7, 8, and 9. The ten digits may be used alone or combined to represent quantities indicating how much, how far, how long, how hot, how expensive, etc. This is the numeral system most commonly used in the United States.

A *place* is the position that a digit occupies and represents the value of the digit. A digit has different values according to the place it occupies. A digit can occupy one of three places, which is the units, tens, or hundreds place. Large Arabic numerals are made easier to read by the use of periods. See Figure 1-2.

A *period* is a group of three places and is separated from the other periods by a comma. The units period (000 through 999) is the first period. The thousands period (1000 through 999,999) is the second period. The millions period (1,000,000 through 999,999,999) is the third period, etc.

The comma is often omitted in numbers that occupy the units place of the thousands period. For example, seven thousand, two hundred twenty-two can be written 7,222 or 7222.

Each period of the digits is read the same way, and the name of that period is put at the end. These period names indicate the location of the digits in regard to the units place.

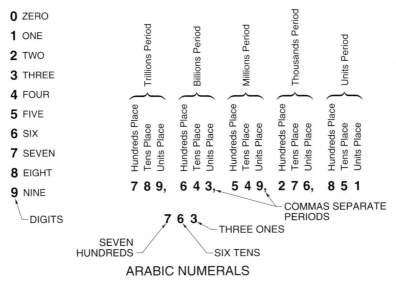

ARABIC NUMERALS

Figure 1-2. Arabic numerals express all numbers by ten digits and combinations of those digits.

For example, 125,000,000,001 is read one hundred twenty-five billion, one. The 125 is in the billions period and the zeros merely fill vacant spaces. Always read each period as if it were alone, then read the next to the right alone, and so on to the end. Do not use the word *and* in reading such numbers. Examples of Arabic numerals and how to read them as word statements are:

10,067,072—Ten million, sixty-seven thousand, seventy-two

1,003,500—One million, three thousand, five hundred

800,010—Eight hundred thousand, ten

87,109—Eighty-seven thousand, one hundred nine

6000—Six thousand

4120—Four thousand, one hundred twenty

302—Three hundred two

PRACTICE PROBLEMS

Arabic Numerals

Write Arabic numerals as word statements, and word statements as Arabic numerals.

 1. 123,507

 2. Seventy-eight million, forty-one thousand, seven
 3. 55,017
 4. 4593
 5. Five hundred six thousand, nine hundred twenty-five
 6. 406,932,672
 7. Thirty-five thousand, thirty-five
 8. One thousand, three
 9. 751,820
 10. Three hundred five thousand, seventy-nine

Roman Numerals

Roman numerals are expressed by the letters I, V, X, L, C, D, and M. While not commonly used in the trades, this system is occasionally used on clock faces, on public buildings such as libraries and museums, and for numbering chapters in books, etc. See Figure 1-3.

When a letter is followed by the same letter, or one lower in value, add the values of the letters. For example, XX = 20 and XV = 15.

When a letter is followed by a letter greater in value, subtract the smaller letter from the larger. For example, IV = 4 and VC = 95.

When a letter is placed between two letters greater in value, subtract the smaller letter from the sum of the other two. For example, XIV = 14 and CVL = 145.

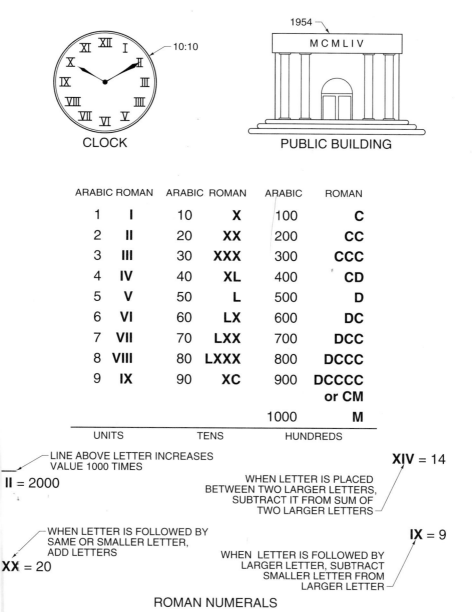

CLOCK

PUBLIC BUILDING

ARABIC	ROMAN	ARABIC	ROMAN	ARABIC	ROMAN
1	I	10	X	100	C
2	II	20	XX	200	CC
3	III	30	XXX	300	CCC
4	IV	40	XL	400	CD
5	V	50	L	500	D
6	VI	60	LX	600	DC
7	VII	70	LXX	700	DCC
8	VIII	80	LXXX	800	DCCC
9	IX	90	XC	900	DCCCC or CM
				1000	M
UNITS		TENS		HUNDREDS	

LINE ABOVE LETTER INCREASES VALUE 1000 TIMES

$\overline{\text{II}}$ = 2000

WHEN LETTER IS PLACED BETWEEN TWO LARGER LETTERS, SUBTRACT IT FROM SUM OF TWO LARGER LETTERS

X**I**V = 14

WHEN LETTER IS FOLLOWED BY SAME OR SMALLER LETTER, ADD LETTERS

XX = 20

WHEN LETTER IS FOLLOWED BY LARGER LETTER, SUBTRACT SMALLER LETTER FROM LARGER LETTER

IX = 9

ROMAN NUMERALS

Figure 1-3. Roman numerals express numbers as letters, instead of digits used in Arabic numerals.

The thousands are represented by the letter M, but a superscript rule (line above) placed over a letter increases the value of the letter a thousand times. For example, the Roman numeral C has a value of 100, but $\overline{C}$ has a value of a thousand times more, or 100,000.

PRACTICE PROBLEMS

Roman Numerals

Write Roman numerals as Arabic numerals and Arabic numerals as Roman numerals.

1. CCLXIX
2. 99
3. 1993
4. LXXI
5. 12
6. $\overline{D}$
7. 144
8. XXXV
9. VM
10. CCXXIV

Adding

Adding is the process of uniting two or more numbers to make one number. Addition is the most common operation in mathematics. The plus sign (+) indicates addition and is used when numbers are added horizontally, or when two numbers are added vertically.

When more than two numbers are added vertically the operation is apparent, and no sign is required. The *sum* is the result of addition. Everyday addition is used to add money. For example, if a person has $17.00 and earns $24.00 more, by adding $17.00 and $24.00, the person has a total of **$41.00**.

Only quantities of the same unit of measure can be added. A *unit of measure* is the specific item(s) being added, subtracted, multiplied, or divided. For example, dollars and gallons (gal.) cannot simply be added. The addition of $205.00 and 110 gal. equals the sum of 315, but the sum is not dollars or gallons.

When adding real items, the answer should contain the unit of measure. For example, 10 nails + 8 nails = 18 nails, or 7 yards (yd) + 5 yd = 12 yd. The unit of measure should be recorded in addition, subtraction, multiplication, and division to avoid many errors in mathematics.

To add whole numbers vertically, place corresponding numbers in aligned columns. Place all the units in each of the numbers in the units column, all the tens in the tens column, all the hundreds in the hundreds column, etc. See Figure 1-4.

Add the columns from top to bottom, beginning with the units column. When the sum of the units column is 0–9, record the sum and add the tens column. When the sum of the units column is greater than 10, record the last digit and carry the remaining digit(s) to the tens column. Follow this procedure for the remaining columns.

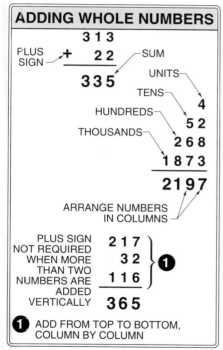

Figure 1-4. Addition is the process of uniting two or more numbers to make one number.

Adding whole numbers horizontally is more difficult than adding vertically. For example, $15 + 120 + 37 + 9 = 181$ shows whole numbers added horizontally. This method is

not as commonly used as the vertical alignment method because mistakes can occur more easily.

Mathematical problems may also be expressed in words. The problem is read, and the operation used to solve the problem is then determined. For example, in a print shop, a press operator prints daily quantities of letterhead of 11,500 on Monday, 12,034 on Tuesday, and 23,564 on Wednesday. Directions are not given to find how much letterhead is printed in the three day period. Addition is used to solve the problem. The total quantity over the three day period is 47,098 ($11,500 + 12,034 + 23,564 = $ **47,098**).

Checking Addition. To check vertically aligned addition problems, add the numbers from bottom to top. To check horizontally aligned addition problems, add the numbers from right to left. The same sum occurs if both operations are performed correctly.

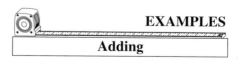
EXAMPLES

Adding

1. Find how many pounds (lb) of copper are stored in four bins containing 4 lb, 52 lb, 268 lb, and 1873 lb.

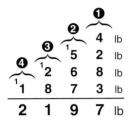

```
            ❶
        ❷  ⌢
      ❸ ⌢  4  lb
   ❹ ⌢ ¹5  2  lb
  ⌢ ¹2  6  8  lb
  ¹1  8  7  3  lb
  ─────────────
  2  1  9  7  lb
```

❶ ADD UNITS COLUMN AND CARRY 1
❷ ADD TENS COLUMN AND CARRY 1
❸ ADD HUNDREDS COLUMN
   AND CARRY 1
❹ ADD THOUSANDS COLUMN

① Add the digits in the units column first. Record the 7 and carry the 1 ten to the tens column.

② Add the tens column including the 1 ten from the units column (1 + 5 + 6 + 7 = 19). Record the 9 and carry the 1 hundred to the hundreds column.

③ Add the hundreds column including the 1 hundred from the tens column (1 + 2 + 8 = 11). Record the 1 and carry the 1 thousand to the thousands column.

④ Add the thousands column including the 1 thousand from the hundreds column (1 + 1 = 2). The sum of the number of pounds of copper in the four bins is **2197 lb**.

2. Find the sum of 25, 20, 19, 12, 32, and 8.

```
      ❷  ❶
     ⌢  ⌢
    ²2  5
     2  0
     1  9
     1  2
     3  2
        8
  ─────────
  1  1  6
```

❶ ADD UNITS COLUMN AND CARRY 2
❷ ADD TENS COLUMN

3. In five months, a drafter works on six projects. The drafter spends 102 hours on Project A, 55 hours on Project B, 190 hours on Project C, 145 hours on Project D, 80 hours on Project E, and 131 hours on Project F. Find how many hours are spent on the six projects.

```
    ❸  ❷  ❶
   ⌢  ⌢  ⌢
  ³1  ¹0  2   PROJECT A
      5  5   PROJECT B
   1  9  0   PROJECT C
   1  4  5   PROJECT D
      8  0   PROJECT E
   1  3  1   PROJECT F
  ─────────────
   7  0  3   TOTAL HOURS
```

❶ ADD UNITS COLUMN AND CARRY 1
❷ ADD TENS COLUMN AND CARRY 3
❸ ADD HUNDREDS COLUMN

4. A surveying party works for six weeks. The first week they survey 61 miles; the second week, 111 miles; the third week, 72 miles; the fourth week, 63 miles; the fifth week, 96 miles; and the sixth week, 48 miles. Find how many miles are surveyed.

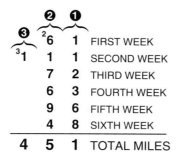

❷	❶	
❸ ²6	1	FIRST WEEK
³1 1	1	SECOND WEEK
7	2	THIRD WEEK
6	3	FOURTH WEEK
9	6	FIFTH WEEK
4	8	SIXTH WEEK
4 5	**1**	TOTAL MILES

❶ ADD UNITS COLUMN AND CARRY 2
❷ ADD TENS COLUMN AND CARRY 3
❸ ADD HUNDREDS COLUMN

5. When purchasing a $16,990.00 truck, the following extras were bought: bed cover, $150.00; set of floor mats, $30.00; spotlight, $55.00; and trailer hitch, $120.00. Find the total cost of the truck.

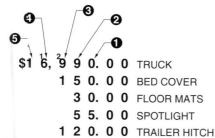

$1 6, 9 9 0. 0 0					TRUCK
1 5 0. 0 0					BED COVER
3 0. 0 0					FLOOR MATS
5 5. 0 0					SPOTLIGHT
1 2 0. 0 0					TRAILER HITCH
$17, 3 4 5. 0 0					TOTAL COST

❶ ADD UNITS COLUMN
❷ ADD TENS COLUMN AND CARRY 2
❸ ADD HUNDREDS COLUMN AND CARRY 1
❹ ADD THOUSANDS COLUMN
❺ ADD TEN THOUSANDS COLUMN

PRACTICE PROBLEMS

Adding

1. 56 + 49 + 17 + 36 + 21
2. 467 + 536 + 84 + 705
3. 8950 + 15,765 + 7732
4. 14,005 + 1204 + 350 + 9786 + 43
5. 12 + 15,045 + 159,056 + 179
6. 144 + 78,045 + 956 + 3563
7. Find the length of the Spacing Gauge.

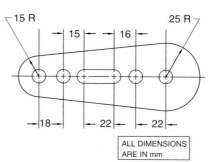

SPACING GAUGE

8. A plumber has five sections of Copper Pipe. Find the total length of pipe.

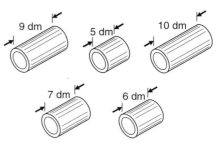

COPPER PIPE

9. A contractor estimates the cost of remodeling a house as follows: $800.00 for masonry, $950.00 for lumber, $110.00 for hardware, $75.00 for trim, $120.00 for paint, and $3500.00 for labor. Find the total estimated cost.

10. Find the total number of watts (W) in Circuit A with all lamps turned on.

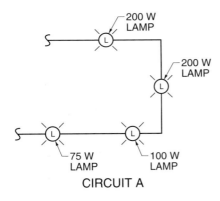

CIRCUIT A

Subtracting

Subtracting is the process of taking one number away from another number (the opposite of adding). In addition numbers are united, while in subtraction one number is taken away from another. The minus sign (–) indicates subtraction. The *minuend* is the number that is subtracted from, and the *subtrahend* is the number to be subtracted. Place the minuend above the subtrahend when vertically aligning numbers

for subtraction. See Figure 1-5. The *difference* is the result of subtraction.

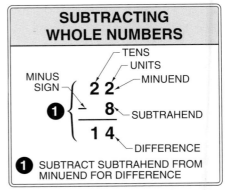

Figure 1-5. Subtraction is the process of taking one number away from another number.

As in addition, the quantities to be subtracted must be of the same unit of measure. For example, pounds and feet cannot simply be subtracted. The subtraction of 35 lb from 110′ has a difference of 75, but the difference is not pounds or feet.

As in addition problems, the first column represents the units column, the second column represents the tens column, etc. Whenever a subtrahend is larger than the corresponding minuend digit, borrow from the next column immediately to the left, and continue the operation.

For example, when subtracting 8 from 22, borrow 1 ten from the tens column, subtract 8 from 12, and record the 4 in the units column. Then record the remaining 1 in the tens column for a difference of 14.

To subtract whole numbers, align the columns and subtract beginning with the units column. To subtract whole numbers in which numbers must be borrowed, align the columns and subtract beginning with the units column. Always borrow from the column to the left.

Checking Subtraction. Addition is used to check subtraction by adding the difference to the subtrahend. The sum of the difference and the subtrahend are the same as the minuend if both operations are performed correctly.

EXAMPLES
Subtracting

1. If 34″ of sheet metal are cut from a piece of stock that is 76″ long, find the remaining stock.

$$
\begin{array}{cc}
\textcircled{2} & \textcircled{1} \\
7 & 6 \\
- \ 3 & 4 \\
\hline
4 & 2
\end{array}
$$

❶ SUBTRACT UNITS COLUMN
❷ SUBTRACT TENS COLUMN

① In the units column take 4 from 6 (6 − 4 = 2). ② Subtract 3 from 7 in the tens column (7 − 3 = 4). This subtraction is commonly understood as 30 from 70 equals 40, which is a common expression when the 4 is set in the tens place. There is **42″** of stock remaining.

2. Subtract 364 from 942.

$$
\begin{array}{cc}
942 & \quad \overset{\textcircled{1}}{9 \ \overset{3}{4} \ (12)} \\
- \ 364 \ = & - \ 3 \quad 6 \quad 4 \\
& \hline
& \qquad \qquad 8
\end{array}
$$

$$
= \quad
\begin{array}{c}
\overset{\textcircled{2}}{\overset{8}{9} \ (13)(12)} \\
- \ 3 \quad 6 \quad 4 \\
\hline
7 \quad 8
\end{array}
$$

$$
= \quad
\begin{array}{c}
\overset{\textcircled{3}}{8 \ (13)(12)} \\
- \ 3 \quad 6 \quad 4 \\
\hline
5 \quad 7 \quad 8
\end{array}
$$

❶ BORROW FROM TENS COLUMN AND SUBTRACT UNITS COLUMN
❷ BORROW FROM HUNDREDS COLUMN AND SUBTRACT TENS COLUMN
❸ SUBTRACT HUNDREDS COLUMN

① Borrow 1 ten from the tens column to get 12 in the units column. Subtract 4 from 12 (12 − 4 = 8). ② Borrow 1 hundred from the hundreds column to get 13 in the tens column. ③ Subtract 6 from 13 (6 − 13 = 7). Subtract 3 from 8 (8 − 3 = 5). The difference is **578**.

PRACTICE PROBLEMS
Subtracting

1. 560,894 − 30,101
2. 467 − 84
3. 8950 − 7732
4. 9786 − 43
5. 159,056 − 9179

6. What is the length of A on the Centering Template?

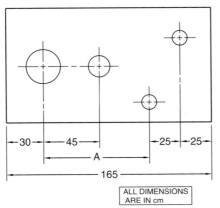

ALL DIMENSIONS
ARE IN cm

CENTERING TEMPLATE

7. A sheet metal worker installs 4″ × 12″ duct in various lengths for a forced air heating system. Find the length of B.

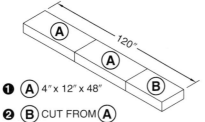

❶ Ⓐ 4″ x 12″ x 48″

❷ Ⓑ CUT FROM Ⓐ

8. The register on a postage machine reads 8090 units on Monday. On Friday it reads 9428 units. Find how many units are used during the week.

9. A tank contains 1200 gal. of water. The tank loses 6 gal. by leakage, while 320 gal. are pumped into it. Find how many gallons are in the Tank.

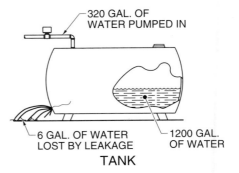

320 GAL. OF WATER PUMPED IN

6 GAL. OF WATER LOST BY LEAKAGE

1200 GAL. OF WATER

TANK

10. A storage shed contains 8579 tons (t) of bar stock from which 3243 t are removed. The shed then receives an additional 4112 t from which 1602 t are removed. Find how many tons remain.

Multiplying

Multiplying is the process of adding one number as many times as there are units in the other. Multiplying is a shortcut for addition. For example, 2 × 6 is the same as 2 taken 6 times, or 2 + 2 + 2 + 2 + 2 + 2. The multiplication or times sign (×) indicates multiplication or multiplied by. See Figure 1-6.

For example, 6 × 5 = 30 is read 6 multiplied by 5 equals 30, or 6 times 5 equals 30. The *multiplicand* is the number being multiplied, and the *multiplier* is the number multiplied by. The *product* is the result of multiplication.

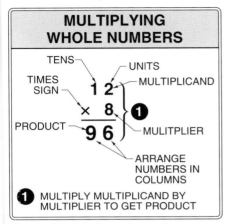

Figure 1-6. Multiplication is the process of adding one number as many times as there are units in the other.

Multiplication tables are used to learn multiplication between different numbers. See Figure 1-7. Learning the table helps recall multiplication of numbers. The larger number is usually used as the multiplicand when the units of measure being multiplied are the same.

1	2	3	4	5	6	7	8	9	10	11	12
MULTIPLICATION TABLE											
2	4	6	8	10	12	14	16	18	20	22	24
3	6	9	12	15	18	21	24	27	30	33	36
4	8	12	16	20	24	28	32	36	40	44	48
5	10	15	20	25	30	35	40	45	50	55	60
6	12	18	24	30	36	42	48	54	60	66	72
7	14	21	28	35	42	49	56	63	70	77	84
8	16	24	32	40	48	56	64	72	80	88	96
9	18	27	36	45	54	63	72	81	90	99	108
10	20	30	40	50	60	70	80	90	100	110	120
11	22	33	44	55	66	77	88	99	110	121	132
12	24	36	48	60	72	84	96	108	120	132	144

Figure 1-7. Multiplication tables are used to help learn and recall multiplication of numbers.

For example, $6' \times 4' = 24'$. If a number to be multiplied represents a unit of measure, identify the unit of measure in the multiplicand, multiplier, and product.

If the multiplier occupies more than the units place, then the product of each figure must be carefully placed in the proper place. The product of the tens place of the multiplier and the multiplicand is written as

1248	multiplicand
× 15	multiplier
6240	product of units place
1248	product of tens place
18,720	final product

The units place in the product of the tens place and multiplicand is empty, and is represented by an empty space. Zero has no value, therefore, any number multiplied by zero equals zero, and zero multiplied by any number equals zero. For example, $0 \times 0 = 0$, $492 \times 0 = 0$, $0 \times 8 = 0$, etc.

When there is a zero in the multiplier, multiply by only those numbers that have value. Place the right-hand figure of each separate product under the figure used to find it. For example, the product of the tens place, when multiplying 568 by 102, is dropped and written as

$$
\begin{array}{r}
568 \\
\times\ 102 \\
\hline
1136 \\
5680 \\
\hline
\mathbf{57,936}
\end{array}
$$

To multiply numbers with zeros to the right-hand side (10, 50, 9000, etc.) the zeros are dropped and the other numbers are multiplied. As many zeros as are in both the multiplicand and multiplier are annexed to (placed after) the product of the other numbers.

For example, when multiplying 270 by 2000 the four zeros are dropped and 27 is multiplied by 2, which equals 54. Annex the four zeros to 54 and the final product equals 540,000.

To multiply by 5, multiply the multiplicand by 10 (annex 0 to the multiplicand), and divide by 2, because 10 divided by 2 equals 5. For example, 438 multiplied by 5 is the same as 4380 divided by 2, which equals 2190.

To multiply by 10, annex 0 to the end of the multiplicand. To multiply by 100, annex 00 to the end of the multiplicand. To multiply by 1000, annex 000 to the end of the multiplicand. For example, 35 multiplied by 1000 is the same as 35 with 000 annexed, which equals 35,000.

To multiply by 25, multiply the multiplicand by 100 (annex 00 to the multiplicand), and divide by 4, because 100 divided by 4 equals 25. For example, 220 multiplied by 25 is the same as 22,000 divided by 4, which equals 5500.

Checking Multiplication. To check multiplication, reverse the multiplier and multiplicand, and perform the operation again. The same product occurs if both operations are performed correctly.

Another method of checking multiplication is to use division. The product is divided by the multiplier, which results as the multiplicand of the original problem.

To multiply whole numbers, align the multiplicand and multiplier vertically. Multiply the multiplicand by the appropriate column of the multiplier. Always place the right-hand figure of each separate product under the figure used to find it.

To multiply whole numbers when the multiplier contains zeros, align the multiplicand and the multiplier vertically. Bring all zeros in the multiplier down and multiply by all other numbers.

EXAMPLES

Multiplying

1. Multiply 346 by 47.

```
      3 4 6
    x   4 7
    2 4 2 2 ← ❶
    1 3 8 4 ← ❷
  1 6, 2 6 2 ← ❸
```

❶ MULTIPLY MULTIPLICAND BY UNITS COLUMN OF MULTIPLIER

❷ MULTIPLY MULTIPLICAND BY TENS COLUMN OF MULTIPLIER

❸ ADD PRODUCTS

① Multiply 346 (multiplicand) by the 7 in the multiplier (346 × 7 = 2422). Record 2422, filling the units, tens, hundreds, and thousands places.

② Multiply 346 by the 4 in the multiplier (346 × 4 = 1384). Record 1384, filling the tens, hundreds, thousands, and ten thousands places. The units place is left blank.

③ Add 2422 and 13,840 (2422 + 13,840 = 16,262). The product of 346 × 47 is **16,262**.

2. Multiply 13,456 by 2004.

```
      1 3, 4 5 6
    x     2 0 0 4
        5 3 8 2 4 ← ❶
    2 6 9 1 2 0 0 ← ❷
  2 6, 9 6 5, 8 2 4 ← ❸
```

❶ MULTIPLY MULTIPLICAND BY UNITS COLUMN OF MULTIPLIER

❷ MULTIPLY MULTIPLICAND BY THOUSANDS COLUMN OF MULTIPLIER

❸ ADD PRODUCTS

① Multiply the multiplicand by 4 and record product (13,456 × 4 = 53,824). ② Bring down the two zeros, plus a blank space to represent the units column. Multiply the multiplicand by 2 (13,456 × 2 = 26912 + 00 + blank units place = 26,912,000). ③ Add products (53,824 + 26,912,000 = 26,965,824). The product of 13,456 × 2004 = **26,965,824**.

PRACTICE PROBLEMS

Multiplying

1. Multiply 144 by 12.

2. Multiply 71 by 35.

3. Multiply 1094 by 18.

4. A screw machine produces 97 units per hour. Find the units produced in 16 hours.

5. A welder fabricates four pieces of Channel Iron. Find the total length of Channel Iron fabricated.

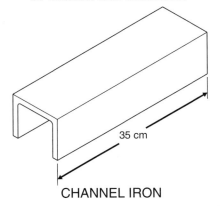

35 cm

CHANNEL IRON

6. Subtract 175 from 5208, and multiply the difference by 97.

7. Find the value of 867 shares of railroad stock at $97.00 a share.
8. If one mile of railroad requires 116 t of steel rail at $290.00 a ton, find the cost of steel needed to construct a road 128 miles in length.
9. A mechanic receives $356.00 for five days work and spends $13.00 a day. Find how much money is left.
10. An electric supplier bought 32 crates of portable radios. Each crate contains 50 radios at a cost of $16.00 a radio. Find the cost of the 32 crates.

Dividing

Dividing is the process of finding how many times one number contains another number. See Figure 1-8. Dividing is the reverse of multiplying. For example, 56 ÷ 8 = 7, or 56 ÷ 7 = 8 is the opposite of 7 × 8 = 56.

The division sign (÷) indicates division. The long division signs also indicate division. See Appendix.

Division may be written with one of the signs, or with one number over another separated with a rule. For example, 6 ÷ 3 = 2 has the same meaning as $\frac{6}{3}$ = 2.

The rule between the numbers indicates division.

The *dividend* is the number to be divided. The *divisor* is the number that the dividend is divided by. The *quotient* is the result of division. The *remainder* is the part of the dividend left over when the quotient is not a whole number.

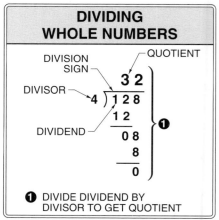

Figure 1-8. Division is the process of finding how many times a number contains another number.

Any remainder is placed over the divisor, and is expressed as a fraction. For example 27 ÷ 4 = 6 ¾. Notice that 4 goes into 27 six times with a remainder of 3. The 3 is placed over the 4, resulting in the fraction ¾.

To divide a number by 10, 100, etc., remove as many places from the right of the dividend as there are zeros in the divisor. For example, 35,900 ÷ 100 = **359**.

To divide by 25, multiply the dividend by 4 and divide the product by 100, which removes two digits from the right-hand side. For example, $450 \div 25$ is the same as $450 \times \frac{4}{100}$, which equals 18.

To record an answer in which the quotient is a whole number, place the quotient over the divisor.

To record an answer in which the quotient is not a whole number, place the remainder over the divisor as a fraction.

To divide a number by 10, 100, etc., cut off as many places from the right of the dividend as there are zeros in the divisor.

To divide a number by 25, multiply the dividend by 4 and divide the product by 100. Cut off two figures from the right.

Checking Division. Multiplication, which is the reverse process of division, is used to check division. For example, to check $655 \div 5 = 131$, multiply 131 (quotient) by 5 (divisor), or $131 \times 5 = 655$. The product of the check is equal to the dividend of the original problem if both operations are performed correctly. If division results in a remainder, then the remainder is added to the product of the check, which results in the dividend of the original problem.

EXAMPLES

Dividing

1. Divide 720 by 5.

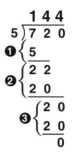

❶ MULTIPLY DIVISOR BY 1 AND SUBTRACT
❷ ANNEX NEXT NUMBER OF DIVIDEND TO DIFFERENCE, MULTIPLY DIVISOR BY 4, AND SUBTRACT
❸ ANNEX NEXT NUMBER OF DIVIDEND TO DIFFERENCE, MULTIPLY DIVISOR BY 4, AND SUBTRACT

The first number of the quotient is the number multiplied by the divisor that gives a product equal or as equal as possible to the first (left-hand) number in the dividend. The number of the quotient cannot be larger than the number in the dividend. The 5 is the divisor, and 7 is the first number. The first number in the quotient is 1 because 1 $\times$ 5 = 5, which is less than 7.

① Multiply 1 by the divisor 5, and subtract the product from 7 (7 − 5 = 2).

② Annex the second number in the dividend to the 2, which results in 22. The number when multiplied by the divisor that equals 22 or near to 22 is the second number of the quotient. That number is 4 because $5 \times 4 = 20$ and $5 \times 5 = 25$. Multiply 4 by the divisor 5, and subtract the product from 22 ($22 - 20 = 2$).

③ Annex the next number of the dividend to that 2, which results in 20. The number that equals 20 or near to 20 when multiplied by the divisor is the next number in the quotient. That number is 4 because $5 \times 4 = 20$. Multiply 4 by the divisor 5 and subtract the product from 20 ($20 - 20 = 0$). There is no remainder. The quotient of $720 \div 5 = \mathbf{144}$.

2. Divide 8936 by 34.

$$34\,\overline{)\,8936}\quad 262\tfrac{28}{34}\text{④}$$

❶ { 6 8

❷ { 2 1 3 / 2 0 4

❸ { 9 6 / 6 8

2 8

❶ MULTIPLY DIVISOR BY 2 AND SUBTRACT
❷ ANNEX NEXT NUMBER OF DIVIDEND TO DIFFERENCE, MULTIPLY DIVISOR BY 6, AND SUBTRACT
❸ ANNEX NEXT NUMBER OF DIVIDEND TO DIFFERENCE, MULTIPLY DIVISOR BY 2, AND SUBTRACT
❹ PLACE DIFFERENCE OVER DIVISOR

① Multiply 2×34 ($2 \times 34 = 68$). Record 68 beneath 89. Subtract 68 from 89 ($89 - 68 = 21$). Record 21.

② Bring the 3 down and multiply 6×34 ($6 \times 34 = 204$). Record 204 beneath 213 and subtract ($213 - 204 = 9$). Record 9.

③ Bring the 6 down and multiply 2×34 ($2 \times 34 = 68$). Record 68 beneath 96. Subtract 68 from 96 ($96 - 68 = 28$).

④ Place 28 over 34. The quotient of $8936 \div 34 = \mathbf{262^{28}\!/_{34}}$.

3. There are 2000 lb in a ton. If a steel mill sells 36,000 lb of steel, find how many tons are sold.

❶ { 2 0 0 0)̶3̶6̶,̶0̶0̶0̶ =

$$2\,\overline{)\,36}\quad 18t$$

❷ { 2 / 1 6

❸ { 1 6

0

❶ CANCEL ZEROS
❷ MULTIPLY DIVISON BY 1 AND SUBTRACT
❸ ANNEX NEXT NUMBER OF DIVIDEND TO DIFFERENCE, MULTIPLY DIVISOR BY 8, AND SUBTRACT

① Cancel all zeros. ② Divide 36 by 2 ($36 \div 2 = 18$). The quotient of $36,000 \div 2000 = \mathbf{18\ t}$.

4. Divide 8700 by 25.

❶
$$\overbrace{8\ 7\ 0\ 0}$$
$$2\ 5\overline{)8\ 7\ 0\ 0} = \text{x} \qquad 4$$
$$3\ 4,8\ \cancel{0}\ \cancel{0}\}❷$$
$$= \textbf{348}$$

❶ MULTIPLY DIVIDEND BY 4
❷ CUT OFF TWO FIGURES FROM RIGHT

① Multiply 8700 × 4 (8700 × 4 = 34,800). ② Cut off two figures from the right to divide the product by 100 (34,800 ÷ 100 = 348). The quotient of 8700 ÷ 25 = **348**.

PRACTICE PROBLEMS

Dividing

1. Divide 1440 by 20.
2. Divide 7200 by 12.
3. Divide 414 by 18.
4. Divide 1656 by 23.
5. If a truck assembly plant assembles 28 trucks per hour, how many hours does it take to assemble 1148 trucks?
6. During one week, an afternoon newspaper sold copies of the paper as follows: Monday, 9462 copies; Tuesday, 10,987 copies; Wednesday, 8455 copies; Thursday, 12,309 copies; Friday, 11,087 copies; Saturday, 15,410 copies. Find how many copies were sold during the six days, and find the average sale per day.

7. A surveyor stakes out the boundaries of six building lots. All of the lots have the same amount of frontage along Oak Street. Find the frontage length of the lots.

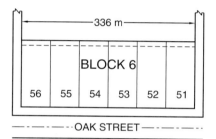

8. If #28 copper conductor costs $11.00 per spool, find how many spools can be bought for $165.00.
9. At the rate of 50 miles per hour (mph), find how long a train takes to travel from Chicago to Salt Lake City.

10. A truck used 230 gal. of diesel fuel during one week. The truck traveled 1705 miles during the week. Find the average miles per gallon.

11. A skid contains 864 books packed in 24 cartons. How many books are in each carton?

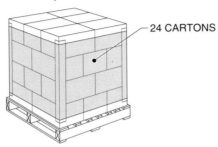

24 CARTONS

12. The total charge for making 15 copies on a library copier is $1.20. What is the cost per copy?

13. A baseball pitcher threw 252 pitches in a nine-inning game. What was the average number of pitches thrown per inning?

14. The fee for traveling 30 miles on a toll road is $1.20. What is the toll per mile?

15. Three pieces of 4′ × 8′ plywood are ripped into 9″ widths. How many 9″ pieces can be ripped from the three pieces of plywood?

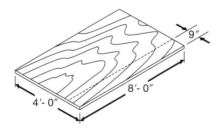

9″

8′- 0″

4′- 0″

16. A fishing team catches four smallmouth bass weighing a total of 16 lb, 8 oz. What is the average weight of the fish?

17. Floor tile for BR2 costs $330.75. What is the cost per square foot?

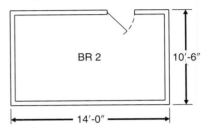

BR 2

10′-6″

14′-0″

18. A front-end loader moves ¾ cubic yard (cu yd) of soil per scoop. How many scoops are required to move 21 cu yd of soil?

19. A total of 420 lamps are assembled by 10 workers in six hours. What is the average number of lamps assembled per hour by each worker?

20. The score of a completed football game is 44–28. What is the average number of points scored per quarter by the winning team?

21. Children in a kindergarten class are the following ages:
Susan—4 yr, 10 mo
John—5 yr, 4 mo
Harold—5 yr, 6 mo
Nancy—5 yr, 2 mo
Mark—5 yr, 0 mo
What is the average age of the children?

Name _____ Date _____

True-False and Completion

T   F   **1.** The subtrahend is the number that is subtracted from.

_____   **2.** A(n)_____ number is any number that cannot be divided by 2 an exact number of times.

_____   **3.** In division, the _____ is the number to be divided.

_____   **4.** _____ is the uniting of two or more numbers to make one number.

T   F   **5.** The multiplicand is the number multiplied by.

_____   **6.** _____ numbers are numbers that have no fractional or decimal parts.

_____   **7.** _____ is the process of finding how many times one number contains another.

_____   **8.** _____ numbers are numbers that can be divided an exact number of times, only by themselves and 1.

T   F   **9.** The number 276,851 is read two hundred seventy-six thousand, and eight hundred fifty-one.

_____   **10.** The _____ is the result obtained from addition.

T   F   **11.** In Roman numerals, a superscript rule over a letter increases the value a hundred times.

T   F   **12.** Arabic numerals use ten digits as symbols.

_____   **13.** The _____ is the result of multiplication.

_____   **14.** In division, the _____ is the part of the dividend left over when the quotient is not a whole number.

T   F   **15.** In Roman numerals, the letter M has a value of 100.

Short Answer

Write as word statements.

1. 501

2. 7808

3. 13,020

4. 734,507,000

Write as Roman numerals.

5. 19

6. 31

7. 40

8. 144

Write as Arabic numerals.

9. Twenty-seven

10. One hundred twenty-three

11. Three thousand, thirty-nine

12. Ten thousand, six

13. XIV

14. DCC

15. $\overline{V}$

16. XXXV

Calculations

_____ 1. Find the sum of 12, 25, 33, 47, 54, 66, 78, and 99.

_____ 2. Find how many double-deck passenger coaches, each carrying 160 people, are needed to carry 1120 people on a commuter train.

_____ 3. An individual paid $121,600.00 for a house and sold it for $124,375.00. Find the profit.

_____ 4. Subtract 5946 from 23,504.

_____ 5. Multiply 4375 by 86.

6. On Monday morning Hirschel Howard and family started on an automobile trip. The odometer registered 2025 miles before starting. On Monday night the odometer registered 2299 miles; Tuesday night, 2558 miles; Wednesday night, 2853 miles; Thursday night, 3145 miles; and Friday night, 3419 miles. Find the average number of miles traveled each day.

7. The owner of a garage had 720 gal. of gasoline at the beginning of the month. During the month the owner bought 1200 gal. and sold 1078 gal. Find the gal. left at the end of the month.

8. An individual sold a house for $91,975.00, a store for $135,150.00, city property for $24,875.00, and received in payment farm land worth $2400.00 per acre (A). Find the acreage of the farm.

9. Divide the difference between 1,236,787 and 54,760 by 3.

10. Multiply 386,789 by 25.

11. Divide 25,600 by 80.

12. Multiply 2000 by 870.

13. Find the sum of 234, 106, 631, 753, 842, and 925.

_____ **14.** A manager of a theater takes in $12,675.00, and pays $3605.00 for pictures, $1825.00 for rent, $2410.00 for wages, and $1335.00 for taxes. Find how much is made in January.

_____ **15.** Denver is 5182′ above sea level. Death Valley is 271′ below sea level. Find how much higher Denver is than Death Valley.

_____ **16.** A used car dealer with
_____ $25,320.00 in cash can buy cars at $1800.00 each. Find the number of cars bought, and how much money is left.

_____ **17.** A truck loaded with coal weighs 14,025 lb. When empty, the truck weighs 4985 lb. Find how much coal is loaded in the truck.

_____ **18.** Multiply the sum of 124, 345, and 567 by 6.

_____ **19.** A factory worker who received $430.00 per week was promoted to a new position paying $2000.00 per month. Using 52 weeks and 12 months as one year, find how much more the worker receives per year in the new position.

_____ **20.** Multiply the difference between 235,904 and 24,409 by the sum of 2, 5, 10, and 18.

Name _____ Date _____

_____ **1.** Multiply 666 by 100.

_____ **2.** Find the sum of 6, 8, 9, 14, 15, 22, 25, 29, 32, 35, and 71.

_____ **3.** Divide 4600 by 25.

_____ **4.** Subtract 184,378 from 2,056,469.

_____ **5.** Multiply 243 by 3200.

_____ **6.** A wire stretches from the top of one pole to the top of another. The top of one pole is 20′ above the surface of a lake, and the top of the other is 6′ below water level. Find how much higher one end of the wire is than the other.

_____ **7.** An electrician cut off two 93′ pieces and four 12′ pieces of conductor from a 1240′ reel. Find the amount of conductor left.

_____ **8.** If a worker's salary was $500.00 per week, and a raise of $1040.00 a year was given, find the salary increase per week (52 weeks per year).

_____ **9.** A contractor paid six laborers $6.00 per hour, 8 hours a day, for five days. Find the total money paid out.

_____ **10.** Each of 40,000 families in a city pays $62.00 for gas, and $80.00 for electricity per month. Find how much money is spent by each family for 12 months.

_____ **11.** A business agent uses a taxi at $60.00 per day. To buy and operate a car costs $10,500.00 per year. Find how much more it costs per day to use a taxi (250 business days per year).

_____ **12.** A welder fabricates 35 steel water tanks for $33,250.00. What is the welding cost per tank?

_____ **13.** In a twenty-story hotel with 56 rooms per floor, 4480 light bulbs are used. If each room contains the same number of bulbs, find how many bulbs are in each room.

_____ **14.** A bakery's roll machine produces 8000 dozen rolls in 8 hours. Find how many single rolls are produced per minute.

_____ **15.** The distance from San Francisco to Salt Lake City is 752 miles; Salt Lake City to Cheyenne, 436 miles; Cheyenne to Omaha, 503 miles; Omaha to Chicago, 461 miles; Chicago to New York, 818 miles. Find how far is it from San Francisco to New York.

FACTORS and CANCELLATION

A factor is a number being multiplied. A prime factor consists of a prime number. An ordinary factor is any factor. Common factors are two or more ordinary factors that divide into two or more numbers an exact number of times.

FACTORS

Whole numbers, also known as integers, are numbers with no fractional or decimal parts. Examples of whole numbers are 1, 2, 3, 4, 5, etc.

An *odd number* is any number that cannot be divided by 2 an exact number of times. Odd numbers always end in 1, 3, 5, 7, or 9. An *even number* is any number that can be divided by 2 an exact number of times. Even numbers always end in 2, 4, 6, 8, or 0.

A *prime number* is any number that can be divided only by itself and 1 an exact number of times. Prime numbers are 1, 2, 3, 5, 7, 11, 13, 17, 19, etc. See Figure 2-1.

The *product* is the result of multiplication. For example, in the problem 12 × 4 = 48, 48 is the product.

WHOLE NUMBERS

1, 2, 3, 4, 5 — ALSO KNOWN AS INTEGERS

WHOLE NUMBERS

1, 3, 5, 7, 9 — CANNOT BE DIVIDED BY 2

ODD NUMBERS

2, 4, 6, 8, 10 — CAN BE DIVIDED BY 2

EVEN NUMBERS

1, 2, 3, 5, 7 — CAN BE DIVIDED BY ITSELF AND 1

PRIME NUMBERS

Figure 2-1. Whole numbers are numbers with no fractional or decimal parts.

27

A *factor* is a number being multiplied. The 12 and 4 are factors of 48. See Figure 2-2.

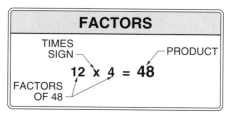

Figure 2-2. Factors are numbers being multiplied.

The factors of a number, when multiplied together, must equal the original number factored. For example, the factors of 8 are 2 and 4. To check this example, the product of these factors is found, or $2 \times 4 = 8$. These factors are correct.

Some numbers have more than two factors. For example, 105 is the product of $3 \times 5 \times 7$. The numbers 3, 5, and 7 are factors of 105. Numbers may have more than one set of factors. The number 105 also has the factors 3 and 35 because $3 \times 35 = 105$. Some numbers have more than two or three factors. For example, $2 \times 2 \times 2 \times 2 = 16$.

Very large numbers can have a great number of factors and may have many different sets of factors. Only one set of factors can be used at one time. Factors may be prime numbers, ordinary numbers, or a combination of both.

Prime Factors

A *prime factor* is a factor consisting of a prime number. For example, 2, 3, 11, 23, etc. are prime factors. The number 1 is a prime number but should not be used as a factor. Also, the numbers for which factors are being found should not be used as factors. See Figure 2-3.

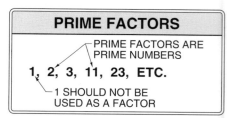

Figure 2-3. Prime factors are factors consisting of prime numbers.

When two or more numbers cannot be divided an exact number of times by any other number, the numbers are prime to each other. For example, 7 and 11 have no other number that divides into them without a remainder. The integers 7 and 11 are prime numbers and are prime to each other.

To find the prime factors of any number, divide by the smallest prime number greater than 1 that divides into the number an exact number of times. A prime number is found through repeated effort, such as by trying to divide by 2 first, then 3, 5, 7, 11, etc. After the divisor is found and the quotient cbtained, the smallest prime number that divides an exact number of times into the quotient is found.

Division is continued until a cuotient that is a prime number is obtained. When a prime number quotient is obtained, the factoring process is completed because a prime number cannot be divided by any number except itself and 1.

The factors for a number include several divisors and the last quotient. Never use 1 or the number being factored as divisors.

To prove any set of factors, find the product of the factors. The product should be the same as the number factored.

① Divide 16 by the smallest prime number greater than 1 that divides into it an exact number of times. The number 2 divides into 16 exactly 8 times.

② Find the smallest prime number that divides into the quotient 8 an exact number of times. The prime factor is 2, leaving a quotient of 4. Divide the quotient 4 by the smallest prime number that divides into 4 an exact number of times. The prime factor is 2, leaving a quotient of 2. The several divisors and the last quotient make up the complete list of factors.

③ The problem is complete because the quotient 2 is a prime number. The prime factors of 16 are **2, 2, 2,** and **2.**

EXAMPLE
Prime Factors

1. Find the prime factors of 16.

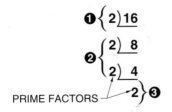

PRIME FACTORS

**PRIME FACTORS
OF 16 = 2, 2, 2, 2**

❶ DIVIDE BY SMALLEST PRIME
   NUMBER GREATER THAN 1
❷ REPEAT AS REQUIRED
❸ QUOTIENT IS A PRIME NUMBER

PRACTICE PROBLEMS
Prime Factors

Find the prime factors of:
1. 78
2. 2772
3. 311
4. 4235
5. 2310

Ordinary Factors

An *ordinary number* is a number that is not a prime number. Ordinary numbers can be prime to each other.

For example, 18 and 25 are not prime numbers, but each can be divided by one or more numbers in addition to itself and 1. The 18 can be divided by 6 and 9 an exact number of times, and the 25 by 5. However, there is no number that divides into *both* 18 and 25 an exact number of times, so they are prime to each other.

An *ordinary factor* is any number whether it is odd, even, or prime. The numbers 2, 3, 4, 5, 6, 7, 8, 9, 15, 18, etc., are all ordinary factors. See Figure 2-4.

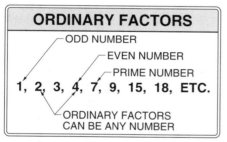

Figure 2-4. Ordinary factors are any factors whether they are odd, even, prime, or prime factors.

The process for finding ordinary factors is similar to the process for finding prime factors. The difference is that in ordinary factoring, any number that divides an exact number of times into the number and quotients can be used. Prime factoring requires that prime numbers are used for factors.

If prime factors are not required, the process of factoring can be simplified by using ordinary factors. In the prime factoring process, only one set of factors can be correct.

In ordinary factoring, several sets of factors can be correct. Several sets of correct ordinary factors can be used because the factoring process starts by using large, medium sized, or small factors.

To find the ordinary factors of a number, select any divisor that divides into the number an exact number of times and find the quotient. Then find the divisor that divides into the quotient an exact number of times.

Division may continue until a quotient is obtained that cannot be divided by any other number except itself and 1. The factors of the number include the divisors and the last quotient. The factors are then prime factors.

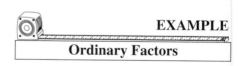

EXAMPLE

Ordinary Factors

1. Factor the number 1716.

The four solutions illustrate that there may be several sets of factors when working with ordinary factors. Ordinary factoring is shorter than prime factoring, because ordinary factoring is not limited to using only prime factors.

SEVERAL SETS OF
FACTORS MAY BE USED

❶ {
6)1716   4)1716   2)1716   13)1716
2) 286   3) 429   6) 858   6) 132
13) 143   13) 143   13) 143   2) 22
11 }❷   11   11   11

6, 2, 13, 11   4, 3, 13, 11   2, 6, 13, 11   13, 6, 2, 11

ORDINARY FACTORS OF 1716

❶ DIVIDE BY ANY DIVISOR
THAT DIVIDES AN EQUAL
NUMBER OF TIMES

❷ QUOTIENT IS A PRIME NUMBER

PRACTICE PROBLEMS

Ordinary Factors

Factor the following numbers. Several solutions are possible.

1. 350
2. 2010
3. 144
4. 260
5. 5390

Common Factors

Common factors are two or more ordinary factors that divide into two or more numbers an exact number of times. See Figure 2-5. For example, 2, 3, 6, and 9 each divide into 18 and 36 an exact number of times. The numbers 2, 3, 6, and 9 are the common factors of 18 and 36. The *highest common factor* is the largest common factor. The number 9 is the largest of the common factors, or the highest common factor, that divides into 18 and 36 an exact number of times.

The distinct difference between common factors and prime and ordinary factors is that prime and ordinary factors relate to one number. Common factors relate to two or more numbers.

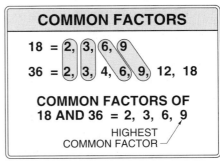

COMMON FACTORS

18 = 2, 3, 6, 9
36 = 2, 3, 4, 6, 9, 12, 18

**COMMON FACTORS OF
18 AND 36 = 2, 3, 6, 9**

HIGHEST
COMMON FACTOR

Figure 2-5. The highest common factor is the largest common factor.

To find the common factors for two or more numbers, begin dividing in order with 2 and then try 3,

4, 5, 6, 7, 8, etc. Continue dividing until all the numbers that divide into the two or more numbers an exact number of times are determined.

A common factor cannot be more than half of the smallest number factored. Only a limited quantity of numbers can be tried. For example, when finding the common factors for the numbers 24 and 60, the number 12 is the highest number tried.

EXAMPLES
Common Factors

1. Find the common factors of 28 and 84.

The number 2 divides into 28 and 84 an exact number of times, so it is a common factor. The number 3 does not divide into 28 and 84 an exact number of times, so it is not a common factor.

The number 4 divides into 28 and 84 an exact number of times, so it is a common factor. The numbers 5 and 6 are not common factors because they do not divide into 28 and 84 an exact number of times.

The number 7 divides into 28 and 84 an exact number of times, so it is a common factor. The numbers 8, 9, 10, 11, 12, 13, and 14 are tried and only 14 divides an exact number of times into 28 and 84. The number 14 is a common factor and the last number tried, because it is half of 28. The common factors of 28 and 84 are **2, 4, 7,** and **14**.

2. Find the common factors of 36, 48, and 72.

The number 2 divides into all three numbers an exact number of times, so it is a common factor. The numbers 3 and 4 divide into all three numbers an exact number of times, so they are common factors.

The number 5 does not divide into all three numbers an exact number of times, so it is not a common factor. The number 6 divides into all three numbers an exact number of times, so it is a common factor.

The numbers 7, 8, 9, 10, and 11 are not common factors. The number 12 is a common factor, but 13, 14, 15, 16, 17, and 18 are not. The number 18 is the last number tried because it is half of 36, the smallest number factored. The common factors of 36, 48, and 72 are **2, 3, 4, 6,** and **12**.

PRACTICE PROBLEMS
Common Factors

Find the common factors.
1. 8, 38, and 44
2. 16, 40, and 96
3. 40, 80, and 100
4. 31, 43, and 57
5. 36, 18, and 54

Find the highest common factor.
6. 16, 32, and 48
7. 12, 64, and 90
8. 36, 72, and 144
9. 11, 33, and 77
10. 35, 85, and 120

CANCELLATION

Cancellation is the process of dividing one factor in the dividend and one factor in the divisor by the same number. Where such division is performed, the number must divide into both factors an exact number of times. See Figure 2-6.

When using cancellation, the terms "find the quotient of" and "cancel" are used interchangeably. For example, find the quotient of $^{96}\!/_{48}$ means the same as cancel $^{96}\!/_{48}$.

The mathematical expression 96 ÷ 48 can also be written $^{96}\!/_{48}$. Using ordinary division and cancellation principles, 96 and 48 can easily be divided and reduced.

To cancel $^{96}\!/_{48}$, find the prime factors of both 96 and 48. The factors of 96 are written as the dividend and the factors of 48 are written as the divisor. A multiplication sign is placed between the factors, because the product of the factors equals the original numbers.

Write the factors of the dividend and the divisor so that one is directly above the other. If the dividend has more factors than the divisor, each factor is indicated in its proper place above the line and a blank is left where a divisor is omitted.

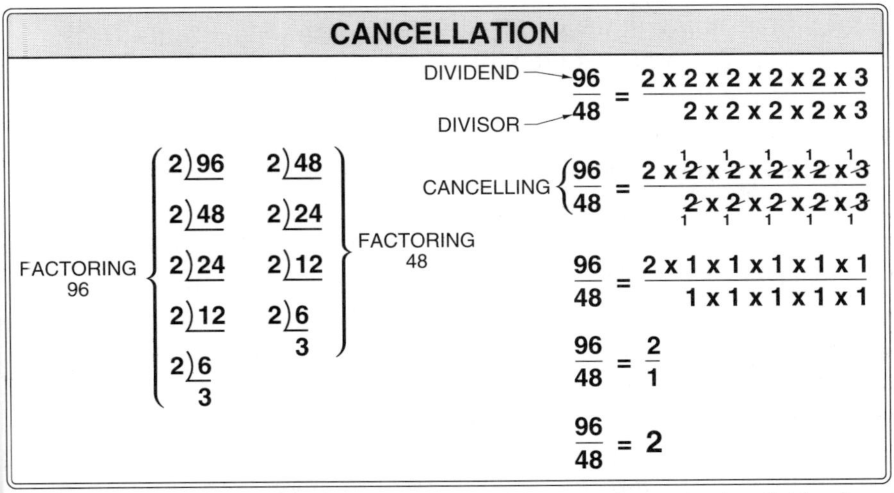

Figure 2-6. Cancellation is the process of dividing one factor in the dividend and one factor in the divisor by the same number.

There are several 2s in both the dividend and the divisor. Select any 2 in the dividend and any 2 in the divisor. Divide both by 2 an exact number of times.

Division is performed four times because there are four 2s in the divisor and five 2s in the dividend. In each case 2 divides into 2 exactly one time. Cross out the 2s and substitute a 1 above and below the crossed out 2s.

There is a 3 in both the dividend and divisor. Both 3s can be divided exactly by 3 one time. Cross out the 3s and substitute a 1 above and below the crossed out 3s. The number 2 is the only number remaining in the dividend.

There are no numbers greater than 1 in the divisor, so dividing is complete. Find the product of the numbers remaining in the dividend and divisor. The products are written as $2/1$, $2 \div 1$, or 2.

Some problems have a group of factors forming a dividend and another group forming the divisor. Cancellation is used to simplify the dividend and divisor. See Figure 2-7.

Cancellation is performed by dividing the dividend and divisor by the same number. The 5 in the dividend and the 10 in the divisor can be divided by 5 an exact number of times. The 5 is crossed out and substituted with a 1. The 10 is crossed

out and is substituted with a 2. The 6 and 3 can be divided by 3 an exact number of times. The 6 is crossed out and substituted with a 2. The 3 is crossed out and substituted with a 1.

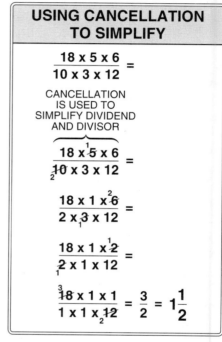

USING CANCELLATION TO SIMPLIFY

$$\frac{18 \times 5 \times 6}{10 \times 3 \times 12} =$$

CANCELLATION IS USED TO SIMPLIFY DIVIDEND AND DIVISOR

Figure 2-7. Cancellation is used to simplify a group of factors divided by another group of factors.

From previous division, there is a 2 in place of the 10. The 2 in the dividend and the 2 in the divisor can be divided by 2 an exact number of times. The 2s are crossed out and substituted with 1s. The 18 and 12 can be divided by 6 an exact number of times. The 18 is crossed out and substituted with a 3. The 12 is crossed out and substituted with a 2. The dividend and divisor of $\dfrac{3 \times 1 \times 1}{1 \times 1 \times 2}$ is $\frac{3}{2}$, $3 \div 2$ which equals **1½**.

There is no sure method of proving cancellation. All problems should be checked to be sure that the same answer is obtained each time.

EXAMPLES

Cancellation

1. Cancel $\dfrac{6 \times 8 \times 12 \times 18 \times 24}{2 \times 3 \times 2 \times 4 \times 6}$.

❶ $\left\{ \dfrac{6 \times 8 \times 12 \times \overset{3}{\cancel{18}} \times 24}{2 \times 3 \times 2 \times 4 \times \underset{1}{\cancel{6}}} \right. =$

❷ $\left\{ \dfrac{6 \times 8 \times \overset{6}{\cancel{12}} \times 3 \times 24}{2 \times 3 \times \underset{1}{\cancel{2}} \times 4 \times 1} \right. =$

❸ $\left\{ \dfrac{6 \times 8 \times 6 \times 3 \times \overset{6}{\cancel{24}}}{2 \times 3 \times 1 \times \underset{1}{\cancel{4}} \times 1} \right. =$

❹ $\left\{ \dfrac{\overset{2}{\cancel{6}} \times 8 \times 6 \times 3 \times 6}{2 \times \underset{1}{\cancel{3}} \times 1 \times 1 \times 1} \right. =$

❺ $\left\{ \dfrac{2 \times \overset{4}{\cancel{8}} \times 6 \times 3 \times 6}{\underset{1}{\cancel{2}} \times 1 \times 1 \times 1 \times 1} \right. =$

❻ $\left\{ \dfrac{2 \times 4 \times 6 \times 3 \times 6}{1 \times 1 \times 1 \times 1 \times 1} \right. = \dfrac{864}{1}$

= 864

❶ DIVIDE 18 AND 6 BY 6
❷ DIVIDE 12 AND 2 BY 2
❸ DIVIDE 24 AND 4 BY 4
❹ DIVIDE 6 AND 3 BY 3
❺ DIVIDE 8 AND 2 BY 2
❻ MULTIPLY FACTORS IN NUMERATOR AND DENOMINATOR AND REDUCE

① The 18 in the dividend and the 6 in the divisor can be divided by 6 an exact number of times. (6 ÷ 6 = 1 and 18 ÷ 6 = 3). The 18 and 6 are crossed out and are substituted with a 3 and 1.

② The 12 in the dividend and the 2 in the divisor can be divided by 2 an exact number of times. The 12 and 2 are crossed out and are substituted with a 6 and 1.

③ The 24 in the dividend and the 4 in the divisor can be divided by 4 an exact number of times. The 24 and 4 are crossed out and are substituted with a 6 and 1.

④ The 6 in the dividend and the 3 in the divisor can be divided by 3 an exact number of times. The 6 and 3 are crossed out and are substituted with a 3 and 1.

⑤ The 8 in the dividend and the 2 in the divisor can be divided by 2 an exact number of times. The 8 and 2 are crossed out and are substituted with a 4 and 1.

⑥ The last factor in the divisor has been cancelled. No more division is possible. When all cancellation is completed, the products of the dividend and divisor are found. The products are 864 for the dividend and 1 for the divisor. The quotient is $^{864}/_{1}$ or 864 ÷ 1 = **864**.

2. Find the quotient of $\dfrac{24 \times 720}{4 \times 12 \times 60}$.

$\mathbf{0} \left\{ \dfrac{\overset{2}{24} \times 720}{4 \times \underset{1}{12} \times 60} = \right.$

$\mathbf{2} \left\{ \dfrac{2 \times \overset{12}{720}}{4 \times 1 \times \underset{1}{60}} = \right.$

$\mathbf{3} \left\{ \dfrac{2 \times \overset{3}{12}}{\underset{1}{4} \times 1 \times 1} = \right.$

$\mathbf{4} \left\{ \dfrac{2 \times 3}{1 \times 1 \times 1} = \dfrac{6}{1} = 6 \right.$

❶ DIVIDE 24 AND 12 BY 12
❷ DIVIDE 720 AND 60 BY 60
❸ DIVIDE 4 AND 12 BY 4
❹ MULTIPLY FACTORS IN NUMERATOR AND DENOMINATOR AND REDUCE

① The 24 in the dividend and the 12 in the divisor can be divided by 12 an exact number of times. The 24 and 12 are crossed out and are substituted with a 2 and 1.

② The 720 in the dividend and the 60 in the divisor can be divided by 60 an exact number of times. The 720 and 60 are crossed out and are substituted with a 12 and 1.

③ The 12 in the dividend and the 4 in the divisor can be divided by 4 an exact number of times. The 12 and 4 are crossed out and are substituted with a 3 and 1. The cancelling is completed.

④ After cancelling, the remaining factors in the dividend are multiplied together, and the remaining factors in the divisor are multiplied together. The product of the divi-dend is 6, and the product of the divisor is 1. The quotient is $\frac{6}{1}$ or $6 \div 1 = \mathbf{6}$.

PRACTICE PROBLEMS

Cancellation

Note: Many different cancellations can be made in the following problems. A variety of cancellation methods will result in the same answer.

Cancel the following.

1. $\dfrac{6 \times 12 \times 20}{4 \times 8 \times 5}$

2. $\dfrac{16 \times 24 \times 18}{36 \times 4 \times 6}$

3. $\dfrac{12 \times 60 \times 36 \times 70}{28 \times 5 \times 48 \times 6}$

4. $\dfrac{16 \times 72 \times 48 \times 32}{8 \times 24 \times 16 \times 36}$

5. $\dfrac{72 \times 70 \times 96 \times 100 \times 150}{25 \times 16 \times 10 \times 36 \times 50}$

6. $\dfrac{64 \times 24 \times 85 \times 32 \times 64}{40 \times 32 \times 64 \times 8 \times 34}$

7. $\dfrac{44 \times 75 \times 63 \times 56 \times 45}{16 \times 35 \times 50 \times 55 \times 18}$

8. $\dfrac{25 \times 16 \times 12}{10 \times 4 \times 6 \times 7}$

9. $\dfrac{120 \times 44 \times 6 \times 7}{72 \times 33 \times 14}$

10. $\dfrac{200 \times 36 \times 30 \times 21}{270 \times 40 \times 15 \times 14}$

FACTORS and CANCELLATION

Review

Name _____ Date _____

True-False and Completion

_____ 1. _____ are two or more ordinary factors that divide into two or more numbers an exact number of times.

T   F   2. There are three sure methods of proving cancellation.

_____ 3. A(n) _____ is any number whether it is odd, even, prime, or a prime factor.

_____ 4. The _____ is the largest common factor.

T   F   5. A factor is a number being multiplied.

T   F   6. To prove factors are correct, add all the factors. The sum should equal the number factored.

_____ 7. _____ is the process of dividing one factor in the dividend and one factor in the divisor by the same number.

_____ 8. A(n) _____ is a factor consisting of a prime number.

T   F   9. When two or more numbers can be divided an exact number of times by any other number, the numbers are prime to each other.

_____ 10. The number _____ is a prime number, but should not be used as a factor.

Calculations

_____ 1. Find the prime factors of 2310.

_____ 2. Factor the number 245.

_____ 3. List the prime numbers between 18 and 36.

4. Cancel $\dfrac{8 \times 6 \times 11 \times 13}{12 \times 22 \times 13 \times 9}$.

5. Find the quotient of $\dfrac{40 \times 48 \times 54 \times 60}{30 \times 24 \times 72 \times 3}$.

6. Find the highest common factor of 18, 36, and 144.

7. Find the prime factors of 539.

8. Factor the number 3550.

9. Factor 73,920 ÷ 10,560 into prime factors and divide by cancellation.

10. Cancel $\dfrac{24 \times 12 \times 48}{6 \times 3 \times 12}$.

11. Find the prime factors of 2508.

12. Cancel $\dfrac{3 \times 8 \times 9 \times 4}{2 \times 27 \times 8}$.

13. Cancel $\dfrac{12 \times 10 \times 18 \times 75}{6 \times 3 \times 5 \times 15}$.

14. Find the highest common factor of 108, 240, and 2350.

15. Factor the number 1024.

16. Find the quotient of $\dfrac{27 \times 33 \times 75 \times 56}{54 \times 11 \times 15 \times 8}$.

17. Find the common factors of 24, 60, and 96.

18. Find the prime factors of 1966.

Name _____ Date _____

_____ **1.** Find the prime factors of 3480.

_____ **2.** Cancel $\dfrac{8 \times 10 \times 24}{5 \times 4 \times 16}$.

_____ **3.** Find the highest common factor of 14, 36, and 78.

_____ **4.** Find the quotient of $\dfrac{21 \times 11 \times 26}{14 \times 13}$.

_____ **5.** Factor the number 1245.

_____ **6.** Find the common factors of 40 and 90.

_____ **7.** Find the quotient of $\dfrac{22 \times 30 \times 36}{5 \times 11 \times 3}$.

_____ **8.** Find the prime factors of 105.

_____ **9.** Factor into prime factors and cancel 2912 ÷ 1456.

_____ **10.** Factor into prime factors and cancel 79,856 ÷ 5704.

_____ **11.** Find the prime factors of 188.

_____ **12.** Cancel $\dfrac{90 \times 66 \times 8}{4 \times 11 \times 30}$.

_____ **13.** Find the highest common factor of 58, 124, and 134.

_____ **14.** Find the common factors of 36, 72, and 96.

_____ **15.** Find the quotient of
$$\frac{66 \times 9 \times 18 \times 5}{22 \times 6 \times 40}.$$

_____ **16.** Find the highest common factor of 12, 26, and 200.

_____ **17.** Cancel $\dfrac{2 \times 4 \times 7}{2 \times 7 \times 2}$.

_____ **18.** Factor the number 936.

_____ **19.** Find the common factors of 24 and 96.

_____ **20.** Find the quotient of
$$\frac{20 \times 12 \times 35}{4 \times 6 \times 7}.$$

_____ **21.** Cancel $\dfrac{3 \times 6 \times 5}{2 \times 4 \times 5}$.

_____ **22.** List the ordinary factors between 7 and 22.

_____ **23.** Find the prime factors of 78.

_____ **24.** Find the common factors of 18, 36, and 54.

_____ **25.** Cancel $^{96}\!/_{48}$.

FRACTIONS:
Adding and Subtracting

A fraction is one part of a whole number. The three main types of fractions are proper fractions, improper fractions, and mixed numbers. A common denominator is the denominator of a group of fractions that has the same denominator.

FRACTIONS

Whole numbers, or integers, are numbers with no fractional or decimal parts. A *fraction* is one part of a whole number. Any number or item smaller than one is a fraction and can be divided into any number of fractional parts. For example, a pipe is a whole, or undivided. See Figure 3-1.

The pipe compares to a whole number, or one. If the pipe is cut into two equal pieces, it is no longer a whole pipe. Each of the two pieces becomes a part of the pipe. One piece of the pipe is one half of the original pipe.

If the pipe is cut into 4 equal pieces, it is no longer a whole pipe. Each of the four pieces becomes a part of the pipe. One piece of the pipe is one fourth of the original pipe.

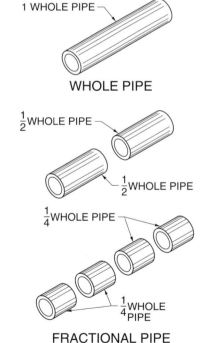

1 WHOLE PIPE

WHOLE PIPE

$\frac{1}{2}$ WHOLE PIPE

$\frac{1}{2}$ WHOLE PIPE

$\frac{1}{4}$ WHOLE PIPE

$\frac{1}{4}$ WHOLE PIPE

FRACTIONAL PIPE

Figure 3-1. A fraction is one part of a whole number.

Fractions are written by placing numbers above and below or on both sides of a fraction bar. Fraction bars are horizontal or inclined. For example, one third is written $\frac{1}{3}$ and one tenth is written $\frac{1}{10}$. See Figure 3-2.

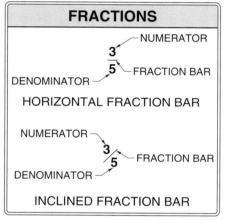

Figure 3-2. The denominator shows into how many parts a whole number is divided. The numerator shows the number of parts taken from the denominator.

The *denominator* shows into how many parts a whole number is divided. The denominator is the lower or right-hand number of a fraction. The *numerator* shows the number of parts taken from the denominator. The numerator is the upper or left-hand number of a fraction. For example, if a pipe is cut into four pieces, and three pieces are used, then three fourths ($\frac{3}{4}$) of the pipe is used. The 4 is the denominator and represents the number of parts into which the pipe

is divided. The 3 is the numerator and represents the number of parts of pipe that were used.

A ruler or tape measure is divided into fractional parts of an inch. The large marks represent inches and are designated by whole numbers. The smaller marks represent fractions of an inch. The largest mark between the 3″ and 4″ marks represents $\frac{1}{2}$″. The largest mark between the 3″ and the $\frac{1}{2}$″ marks represents $\frac{1}{4}$″. The largest mark between the 3″ and the $\frac{1}{4}$″ marks represents $\frac{1}{8}$″. See Figure 3-3.

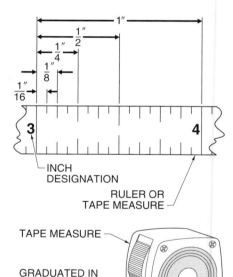

Figure 3-3. A ruler or tape measure is divided into fractional parts of an inch.

Rulers and tape measures may also be graduated (divided and marked off) by $\frac{1}{16}''$, $\frac{1}{32}''$, and $\frac{1}{64}''$. Rulers and tape measures with graduation are necessary for work demanding closer accuracy.

A circle is one whole and can be divided into fractional parts. See Figure 3-4. A circle can be divided into two equal parts, or halves. If the halves are divided into two equal parts, then the circle is divided into fourths.

One half equals two fourths and is written $\frac{1}{2} = \frac{2}{4}$. If the fourths are divided into two equal parts, then the circle is divided into eighths. One fourth equals two eighths and is written $\frac{1}{4} = \frac{2}{8}$.

If the eighths are divided into two equal parts, then the circle is divided into sixteenths. One eighth equals two sixteenths and is written $\frac{1}{8} = \frac{2}{16}$.

PRACTICE PROBLEMS

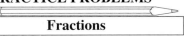

Fractions

A steel rod is cut into 18 equal parts.
1. Find the fraction that represents one of these parts.
2. Find the fraction that represents two of these parts.

A pipe is cut into 6 equal parts.
3. Find the fraction that represents one of these parts.
4. Find the fraction that represents two of these parts.
5. Find the fraction that represents three of these parts.

Types of Fractions

Fractions may be proper fractions, improper fractions, or mixed numbers. Proper fractions are less than one. Improper fractions and mixed numbers are more than one. See Figure 3-5.

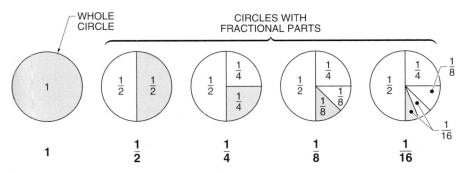

Figure 3-4. A circle can be divided into fractional parts.

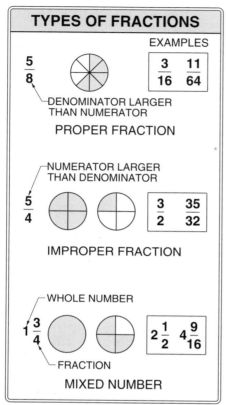

TYPES OF FRACTIONS

EXAMPLES

$\frac{5}{8}$ — DENOMINATOR LARGER THAN NUMERATOR

$\frac{3}{16}$ $\frac{11}{64}$

PROPER FRACTION

NUMERATOR LARGER THAN DENOMINATOR — $\frac{5}{4}$

$\frac{3}{2}$ $\frac{35}{32}$

IMPROPER FRACTION

WHOLE NUMBER — $1\frac{3}{4}$ — FRACTION

$2\frac{1}{2}$ $4\frac{9}{16}$

MIXED NUMBER

Figure 3-5. Fractions may be proper fractions, improper fractions, or mixed numbers.

Proper Fractions. A *proper fraction* is a fraction with the denominator larger than the numerator. For example, $\frac{1}{8}$, $\frac{1}{4}$, $\frac{1}{2}$, $\frac{9}{16}$, $\frac{25}{32}$, $\frac{63}{64}$, etc. are proper fractions.

Improper Fractions. An *improper fraction* is a fraction with the numerator larger than the denominator. For example, $\frac{7}{5}$, $\frac{8}{3}$, $\frac{4}{3}$, $\frac{10}{4}$, $\frac{87}{15}$, $\frac{21}{7}$, $\frac{8}{7}$, $\frac{15}{4}$, etc. are improper fractions.

Mixed Numbers. A *mixed number* is a combination of a whole number and a proper fraction. For example, $1\frac{1}{8}$, $2\frac{1}{4}$, $4\frac{1}{2}$, $6\frac{9}{16}$, $8\frac{25}{32}$, $11\frac{63}{64}$, etc. are mixed numbers.

Mixed numbers and improper fractions are closely related. If one is known, the other can be calculated. Mixed numbers can be changed to improper fractions. Improper fractions can be changed to whole numbers or mixed numbers. See Figure 3-6.

To change a mixed number to an improper fraction, multiply the whole number by the denominator and then add the numerator. Place the new numerator above the denominator.

For example, to change $2\frac{3}{4}$ to an improper fraction, multiply 2 by 4 ($2 \times 4 = 8$) and add 3 ($8 + 3 = 11$). Place the 11 above 4, the original denominator ($\frac{11}{4}$).

To change an improper fraction to a whole number or a mixed number, divide the numerator by the denominator. In a fraction, the horizontal line between the numerator and the denominator indicates division, just the same as 10 ÷ 5 indicates division. The fraction $\frac{10}{5}$ means that 10 (numerator) is divided by 5 (denominator).

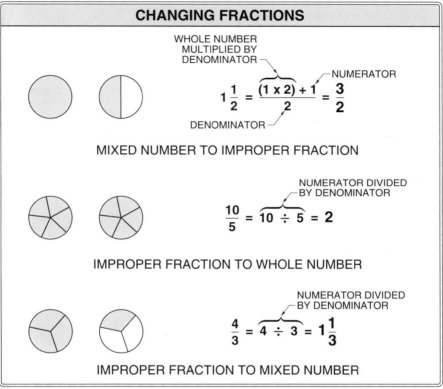

Figure 3-6. Mixed numbers can be changed to improper fractions, and improper fractions can be changed to whole or mixed numbers.

For example, the fraction $\frac{8}{8}$ is an object divided into 8 equal parts and 8 parts are indicated. If an object is divided into 8 parts and 8 parts are indicated, the whole object is indicated because $\frac{8}{8} = \mathbf{1}$.

If an object is divided into 10 parts and 20 parts are indicated, then two objects divided into 10 parts are indicated because $\frac{20}{10} = \mathbf{2}$. Improper fractions can result in whole numbers.

Improper fractions can also result in mixed numbers. For example, if the numerator is divided by the denominator in the fraction $\frac{10}{3}$, the quotient is 3 with a remainder of 1. The remainder is used as a numerator over the original denominator ($\frac{10}{3} = \mathbf{3\frac{1}{3}}$).

To check the answer after changing a mixed number to an improper fraction, change the answer to a mixed number. The check should result in the original mixed number.

To check the answer of changing an improper fraction to a mixed number, change the answer to an improper fraction. The check should result in the original improper fraction.

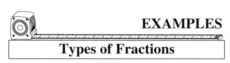

EXAMPLES

Types of Fractions

1. Change $6\frac{8}{9}$ to an improper fraction.

WHOLE NUMBER
NUMERATOR
❶
$$6\frac{8}{9} = \frac{(6 \times 9) + 8}{9} = \frac{62}{9}$$
DENOMINATOR

❶ MULTIPLY WHOLE NUMBER BY DENOMINATOR, ADD NUMERATOR, AND PLACE SUM OVER DENOMINATOR

① Multiply 6 by 9 (6 × 9 = 54). Add 8 to 54 (54 + 8 = 62). Place 62 over 9 ($\frac{62}{9}$ = $\frac{62}{9}$).

2. Change $25\frac{2}{3}$ to an improper fraction.

❶
$$25\frac{2}{3} = \frac{(25 \times 3) + 2}{3} = \frac{77}{3}$$

❶ MULTIPLY WHOLE NUMBER BY DENOMINATOR, ADD NUMERATOR, AND PLACE SUM OVER DENOMINATOR

① Multiply 25 by 3 (25 × 3 = 75). Add 2 to 75 (75 + 2 = 77). Place 77 over 3 ($25\frac{2}{3}$ = $\frac{77}{3}$).

3. Change $\frac{5}{3}$ to a mixed number.

$$\frac{5}{3} = 3\overline{)5} = 1\frac{2}{3}$$
$$\frac{3}{2}$$

❶ DIVIDE NUMERATOR BY DENOMINATOR

① Divide 5 by 3 (5 ÷ 3 = 1 with a remainder of 2). The 1 is the whole number of the mixed number and the fraction is 2 over 3 ($\frac{5}{3}$ = $1\frac{2}{3}$).

4. Change $\frac{75}{20}$ to a mixed number.

$$\frac{75}{20} = 20\overline{)75} = 3\frac{15}{20} = 3\frac{3}{4}$$
$$\frac{60}{15}$$

❶ DIVIDE NUMERATOR BY DENOMINATOR
❷ REDUCE AS REQUIRED

① Divide 75 by 20 (75 ÷ 20 = 3 with a remainder of 15). The 3 is the whole number of the mixed number, and the fraction is 15 over 20. ② Reduce the fraction to lowest terms ($\frac{75}{20}$ = $3\frac{15}{20}$ = $3\frac{3}{4}$).

PRACTICE PROBLEMS

Types of Fractions

Change mixed numbers to improper fractions and improper fractions to mixed numbers.

1. $8^3/_7$
2. $12^5/_9$
3. $20^1/_{11}$
4. $^8/_5$
5. $83^1/_3$
6. $^{125}/_2$
7. $46^7/_{10}$
8. $^{94}/_7$
9. $^{763}/_{18}$
10. $23^{75}/_{76}$

Reducing Fractions

Reducing fractions is changing a fraction from one fractional form to another. Fractions are generally considered to be reduced when they are changed to lower terms. Fractions are generally considered to be changed when they are changed to higher terms. The *terms* of the fraction are the numerator and denominator. For example, the terms of the fraction $^3/_4$ are 3 and 4.

Changing to Higher Terms.
Changing to higher terms is multiplying the numerator and denomi-

nator by the same number. For example, the fraction $^1/_3$ can be changed to higher terms by multiplying 1 (numerator) and 3 (denominator) by 2 ($\dfrac{1 \times 2}{3 \times 2}$ = $^2/_6$). See Figure 3-7. When a fraction is changed to higher terms, its value does not change. The fraction $^4/_6$ has exactly the same value as $^2/_3$.

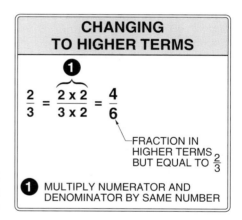

Figure 3-7. To change a fraction to higher terms, multiply the numerator and denominator by the same number.

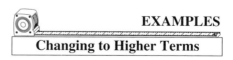

EXAMPLES

Changing to Higher Terms

1. Change $^1/_9$ to higher terms by multiplying the terms by 2.

NUMERATOR — **❶** — SAME NUMBER

$$\frac{1}{9} = \frac{\overbrace{1 \times 2}}{9 \times 2} = \frac{2}{18}$$

— DENOMINATOR

❶ MULTIPLY NUMERATOR AND DENOMINATOR BY SAME NUMBER

① Multiply 1 and 9 by 2 $(\frac{1 \times 2}{9 \times 2} = \frac{2}{18})$.

2. Change $\frac{1}{16}$ to higher terms by multiplying the terms by 2.

$$\frac{1}{16} = \frac{\overbrace{1 \times 2}}{16 \times 2} = \frac{2}{32}$$

❶ MULTIPLY NUMERATOR AND DENOMINATOR BY SAME NUMBER

① Multiply 1 and 16 by 2 $(\frac{1 \times 2}{16 \times 2} = \frac{2}{32})$.

PRACTICE PROBLEMS

Changing to Higher Terms

Change to higher terms by multiplying the terms by 2.

1. $\frac{1}{3}$
2. $\frac{1}{4}$
3. $\frac{1}{13}$
4. $\frac{1}{32}$
5. $\frac{1}{20}$

Changing to Higher Terms with a Given Denominator. To change a fraction to higher terms with a given denominator, divide the new denominator by the denominator of the fraction. Then multiply the numerator and denominator of the fraction by the quotient. See Figure 3-8.

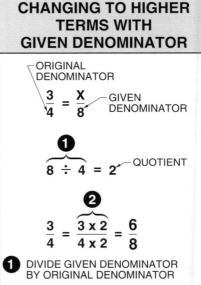

CHANGING TO HIGHER TERMS WITH GIVEN DENOMINATOR

ORIGINAL DENOMINATOR

$$\frac{3}{4} = \frac{X}{8}$$ — GIVEN DENOMINATOR

❶

$$8 \div 4 = 2$$ — QUOTIENT

❷

$$\frac{3}{4} = \frac{3 \times 2}{4 \times 2} = \frac{6}{8}$$

❶ DIVIDE GIVEN DENOMINATOR BY ORIGINAL DENOMINATOR

❷ MULTIPLY NUMERATOR AND DENOMINATOR BY QUOTIENT

Figure 3-8. To change a fraction to higher terms with a given denominator, divide the new denominator by the denominator of the fraction, and multiply the numerator and denominator by the quotient.

For example, to change $\frac{1}{2}$ to higher terms to make the new denominator 8, divide 8 by 2 (8 ÷ 2 = 4). Multiply 4 by the terms 1 and 2 $(\frac{4 \times 1}{4 \times 2} = \frac{4}{8})$. The fraction $\frac{1}{4}$ equals $\frac{2}{8}$, and $\frac{2}{4}$ equals $\frac{1}{2}$, so $\frac{1}{2}$ equals $\frac{4}{8}$.

EXAMPLES

Changing to Higher Terms with a Given Denominator

1. Change $\frac{3}{9}$ to twenty-sevenths.

ORIGINAL DENOMINATOR

$$\frac{3}{9} = \frac{X}{27}$$ GIVEN DENOMINATOR

❶
$$27 \div 9 = 3$$ QUOTIENT

❷
$$\frac{3}{9} = \frac{3 \times 3}{9 \times 3} = \frac{9}{27}$$

❶ DIVIDE GIVEN DENOMINATOR BY ORIGINAL DENOMINATOR

❷ MULTIPLY NUMERATOR AND DENOMINATOR BY QUOTIENT

To change $\frac{3}{9}$ to higher terms to make the new denominator 27, ① divide 27 by 9 (27 ÷ 9 = 3). ② Multiply 3 by the terms 3 and 9 ($\frac{3 \times 3}{9 \times 3}$ = $\frac{9}{27}$).

2. Change $\frac{2}{3}$ to thirty-ninths.

$$\frac{2}{3} = \frac{X}{39}$$

❶
$$39 \div 3 = 13$$

❷
$$\frac{2}{3} = \frac{2 \times 13}{3 \times 13} = \frac{26}{39}$$

❶ DIVIDE GIVEN DENOMINATOR BY ORIGINAL DENOMINATOR

❷ MULTIPLY NUMERATOR AND DENOMINATOR BY QUOTIENT

To change $\frac{2}{3}$ to higher terms to make the new denominator 39, ① divide 39 by 3 (39 ÷ 3 = 13). ② Multiply 13 by the terms 2 and 3 ($\frac{13 \times 2}{13 \times 3}$ = $\frac{26}{39}$).

PRACTICE PROBLEMS

Changing to Higher Terms with a Given Denominator

1. Change $\frac{5}{6}$ to sixtieths.
2. Change $\frac{15}{16}$ to sixty-fourths.
3. Change $\frac{4}{5}$ to fortieths.
4. Change $\frac{9}{25}$ to hundredths.
5. Change $\frac{7}{12}$ to forty-eighths.
6. Change $\frac{9}{16}$ to eightieths.
7. Change $\frac{14}{15}$ to ninetieths.
8. Change $\frac{7}{8}$ to fifty-sixths.
9. Change $\frac{3}{11}$ to seventy-sevenths.
10. Change $\frac{10}{12}$ to seventy-seconds.

Reducing to Lower Terms. To reduce a fraction to lower terms, divide the numerator and denominator by the same number. This process is the reverse of changing a fraction to higher terms. See Figure 3-9.

Fractions are easier to work with when reduced to lower terms. In many cases, there are a number of lower terms to which a fraction can be reduced. Fractions should always be reduced to the lowest terms possible.

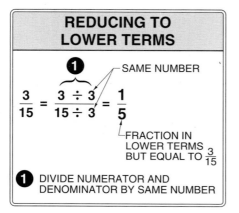

Figure 3-9. To reduce a fraction to lower terms, divide the numerator and denominator by the same number.

For example, the fraction $\frac{3}{9}$ can be reduced to lower terms by dividing 3 (numerator) and 9 (denominator) by the same number ($\frac{3 \div 3}{9 \div 3} = \frac{1}{3}$). Reducing $\frac{3}{9}$ to lower terms does not change its value.

 **EXAMPLES**

Reducing to Lower Terms

1. Reduce $\frac{10}{20}$ to lower terms.

❶
$$\frac{10}{20} = \frac{10 \div 5}{20 \div 5} = \frac{2}{4}$$

❶ DIVIDE NUMERATOR AND DENOMINATOR BY 5

❷
$$\frac{10}{20} = \frac{10 \div 2}{20 \div 2} = \frac{5}{10}$$

❷ DIVIDE NUMERATOR AND DENOMINATOR BY 2

❸
$$\frac{10}{20} = \frac{10 \div 10}{20 \div 10} = \frac{1}{2}$$

❸ DIVIDE NUMERATOR AND DENOMINATOR BY 10

① Divide 5 into the terms 10 and 20 ($\frac{10 \div 5}{20 \div 5} = \frac{2}{4}$).

② Divide 2 into the terms 10 and 20 ($\frac{10 \div 2}{20 \div 2} = \frac{5}{10}$). The fraction $\frac{5}{10}$ is also correct because it has the same value as $\frac{10}{20}$ and is in lower terms.

③ Divide 10 into the terms 10 and 20 ($\frac{10 \div 10}{20 \div 10} = \frac{1}{2}$). The fraction $\frac{1}{2}$ is also correct because it has the same value as $\frac{10}{20}$ and is in lower terms.

In some cases more than one answer is possible when reducing to lower terms. All answers are equally correct and have the same value. The answers $\frac{2}{4}$, $\frac{5}{10}$, and $\frac{1}{2}$ are all correct and have the same value as $\frac{10}{20}$.

To check answers, multiply the terms of the answer by the same number by which they were divided. For example, for the answer $\frac{1}{2}$ the terms 1 and 2 are multiplied by 10 ($\frac{1 \times 10}{2 \times 10} = \frac{10}{20}$). This proves that $\frac{1}{2}$ is a reduced lower term of $\frac{10}{20}$.

2. Reduce $\frac{5}{15}$ to lower terms.

$$\frac{5}{15} = \overbrace{\frac{5 \div 5}{15 \div 5}}^{\textbf{①}} = \frac{1}{3}$$

① DIVIDE NUMERATOR AND DENOMINATOR BY 5

① Divide 5 into the terms 5 and 15 ($\frac{5 \div 5}{15 \div 5} = \frac{1}{3}$). This problem has no other possible answers.

PRACTICE PROBLEMS

Reducing to Lower Terms

Reduce to lower terms.
1. $\frac{11}{33}$
2. $\frac{4}{12}$
3. $\frac{8}{22}$
4. $\frac{15}{55}$
5. $\frac{14}{49}$

Reducing to Lower Terms with a Given Denominator. To reduce a fraction to lower terms with a given denominator, divide the denominator of the fraction by the given denominator. Then divide the numerator and the denominator of the fraction by the quotient. See Figure 3-10.

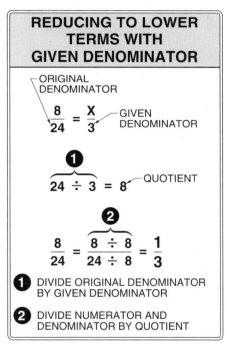

REDUCING TO LOWER TERMS WITH GIVEN DENOMINATOR

$$\frac{8}{24} = \frac{X}{3}$$

(ORIGINAL DENOMINATOR, GIVEN DENOMINATOR)

①

$$24 \div 3 = 8 \quad \text{(QUOTIENT)}$$

②

$$\frac{8}{24} = \frac{8 \div 8}{24 \div 8} = \frac{1}{3}$$

① DIVIDE ORIGINAL DENOMINATOR BY GIVEN DENOMINATOR

② DIVIDE NUMERATOR AND DENOMINATOR BY QUOTIENT

Figure 3-10. To reduce a fraction to lower terms with a given denominator, divide the denominator of the fraction by the given denominator. Then divide the numerator and the denominator of the fraction by the quotient.

For example, to reduce the fraction $\frac{12}{16}$ to fourths, divide the 16 (denominator) by 4 ($16 \div 4 = \textbf{4}$).

Divide 4 into 12 (numerator) and 16 ($\frac{12 \div 4}{16 \div 4}$ = ¾). Reducing to ¾ does not change its value.

EXAMPLE

Reducing to Lower Terms with a Given Denominator

1. Reduce ⁸⁄₃₂ to fourths.

$$\underset{\substack{\text{ORIGINAL} \\ \text{DENOMINATOR}}}{\overset{}{\frac{8}{32}}} = \underset{\substack{\text{GIVEN} \\ \text{DENOMINATOR}}}{\frac{X}{4}}$$

❶
$$32 \div 4 = 8 \quad \text{QUOTIENT}$$

❷
$$\frac{8}{32} = \frac{8 \div 8}{32 \div 8} = \frac{1}{4}$$

❶ DIVIDE ORIGINAL DENOMINATOR BY GIVEN DENOMINATOR

❷ DIVIDE NUMERATOR AND DENOMINATOR BY QUOTIENT

① Divide 32 (denominator) by 4 (32 ÷ 4 = 8). ② Divide 8 into 8 (numerator) and 32 ($\frac{8 \div 8}{32 \div 8}$ = ¼).

PRACTICE PROBLEMS

Reducing to Lower Terms with a Given Denominator

1. Reduce ¹⁵⁄₂₀ to fourths.
2. Reduce ³⁶⁄₄₀ to tenths.
3. Reduce ²⁴⁄₃₆ to sixths.
4. Reduce ⁵⁰⁄₇₅ to thirds.
5. Reduce ¹²⁄₃₆ to ninths.
6. Reduce ¹⁶⁄₂₀ to fifths.
7. Reduce ²⁰⁄₄₄ to elevenths.
8. Reduce ³⁰⁄₄₅ to fifteenths.
9. Reduce ¹⁶⁄₇₆ to nineteenths.
10. Reduce ²⁵⁄₁₁₅ to twenty-thirds.
11. Reduce ¹²⁄₄₈ to fourths.
12. Reduce ²⁴⁄₆₀ to fifths.

Reducing to Lowest Possible Terms. This process is similar to reducing fractions to lower terms. However, when reducing to lower terms, reducing to lowest possible terms is not required. In all mathematical expressions having a fraction for the answer, the fraction should be reduced to lowest possible terms.

Cancellation is used to reduce a fraction to its lowest possible terms. The fraction is reduced until no further reduction is possible.

When only a 1 can be divided into the numerator and denominator of the fraction, the fraction is reduced to its lowest possible terms. See Figure 3-11.

For example, to reduce the fraction ²⁴⁄₃₀ to its lowest possible terms, cancel 24 and 30 by dividing by 2 ($\frac{24 \div 2}{30 \div 2}$ = ¹²⁄₁₅). The 2 does not divide into both 12 and 15 an exact number of times, so 3 (the next higher number that will divide into 12 and 15 an exact number of times) is used.

Cancel 12 and 15 by dividing by 3 ($\frac{12 \div 3}{15 \div 3}$ = ⁴⁄₅). No number other than 1 divides into 4 and 5 an exact number of times, so the lowest possible terms are ⁴⁄₅.

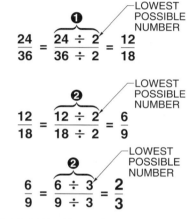

EXAMPLES

**Reducing to
Lowest Possible Terms**

1. Reduce ²⁴⁄₃₆ to its lowest possible terms.

$$\frac{24}{36} = \overbrace{\frac{24 \div 2}{36 \div 2}}^{\text{❶}} = \frac{12}{18} \quad \text{—LOWEST POSSIBLE NUMBER}$$

$$\frac{12}{18} = \overbrace{\frac{12 \div 2}{18 \div 2}}^{\text{❷}} = \frac{6}{9} \quad \text{—LOWEST POSSIBLE NUMBER}$$

$$\frac{6}{9} = \overbrace{\frac{6 \div 3}{9 \div 3}}^{\text{❷}} = \frac{2}{3} \quad \text{—LOWEST POSSIBLE NUMBER}$$

❶ DIVIDE NUMERATOR AND DENOMINATOR BY LOWEST POSSIBLE NUMBER

❷ REPEAT AS REQUIRED

① Cancel 24 and 36 by dividing by 2 ($\frac{24 \div 2}{36 \div 2}$ = ¹²⁄₁₈). ② Cancel 12 and 18 by dividing by 2 ($\frac{12 \div 2}{18 \div 2}$ = ⁶⁄₉). The 2 does not divide into both 6 and 9 an exact number of times, so 3 (next higher number) is used. Cancel 6 and 9 by dividing by 3 ($\frac{6 \div 3}{9 \div 3}$ = ²⁄₃). No further cancellation is possible.

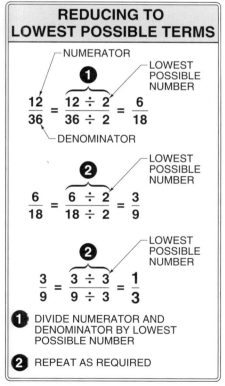

Figure 3-11. To reduce a fraction to the lowest possible terms, divide the numerator and denominator by the lowest possible number.

2. Reduce $^{242}/_{528}$ to lowest terms.

❶

$$\frac{242}{528} = \frac{242 \div 2}{528 \div 2} = \frac{121}{264}$$

❷

$$\frac{121}{264} = \frac{121 \div 11}{264 \div 11} = \frac{11}{24}$$

❶ DIVIDE NUMERATOR AND
DENOMINATOR BY LOWEST
POSSIBLE NUMBER

❷ REPEAT AS REQUIRED

The fraction $^{242}/_{528}$ has large numbers for the numerator and denominator. Reducing may be more difficult when using large numbers for cancellation than when using small numbers. Therefore, cancel using small numbers first. Begin with 2 and determine whether it divides exactly into both the numerator and denominator.

① Cancel 242 and 528 by dividing by 2 ($\frac{242 \div 2}{528 \div 2} = {}^{121}/_{264}$). The 2 does not divide into 121 and 264 an exact number of times. The numbers 3, 4, 5, 6, 7, 8, 9, and 10 do not divide into 121 and 264 an exact number of times. The lowest number that will divide into 121 and 264 an exact number of times is 11.

② Cancel 121 and 264 by dividing by 11 ($\frac{121 \div 11}{264 \div 11} = {}^{11}/_{24}$). The number

11 is the last number tried because no other number except 1 divides exactly into the numerator 11.

PRACTICE PROBLEMS

**Reducing to
Lowest Possible Terms**

Reduce to lowest possible terms.

1. $^{121}/_{165}$
2. $^{144}/_{156}$
3. $^{125}/_{320}$
4. $^{1728}/_{864}$
5. $^{55}/_{121}$
6. $^{640}/_{1280}$
7. $^{120}/_{960}$
8. $^{119}/_{161}$
9. $^{143}/_{187}$
10. $^{2236}/_{8944}$

Common Denominator

Reducing or changing fractions from one form to another, such as reducing to lower terms, changing to higher terms, and changing improper fractions to mixed numbers, deals with a single fraction. Fractions that are added and subtracted are considered a group of fractions.

The denominator is the lower portion of a fraction and shows into how many parts a unit is divided. If two or more fractions have the same denominator, the fractions have a common denominator. A *common denominator* is the denominator of a group of fractions that has the same denominator. *Similar fractions* are fractions that have a common denominator.

For example, $1/8$, $2/8$, $3/8$, $4/8$, $5/8$, $6/8$, and $7/8$ are similar fractions because they have a common denominator of 8. Before addition and subtraction of fractions is performed, certain preparations are necessary, such as finding the lowest common denominator (LCD) of the fractions to be added or subtracted.

Lowest Common Denominator (LCD). The *lowest common denominator* is the smallest number into which the denominators of a group of two or more fractions divides an exact number of times. The group of fractions $4/10$, $3/8$, $3/10$, $6/7$, and $2/3$ does not have a common denominator because there are four different denominators. The LCD must be found to add or subtract these fractions.

To find the LCD for a group of fractions, arrange the denominators in a horizontal row. A division sign is drawn under the row. The smallest number that divides an exact number of times into two or more of the denominators is found.

Divide and place the quotients under the denominators that are divisible an exact number of times. The denominators that are not divisible are brought down. Repeat these steps until division is no longer possible. Multiply all the divisors and quotients (numbers left in the last line). The product is the LCD. See Figure 3-12.

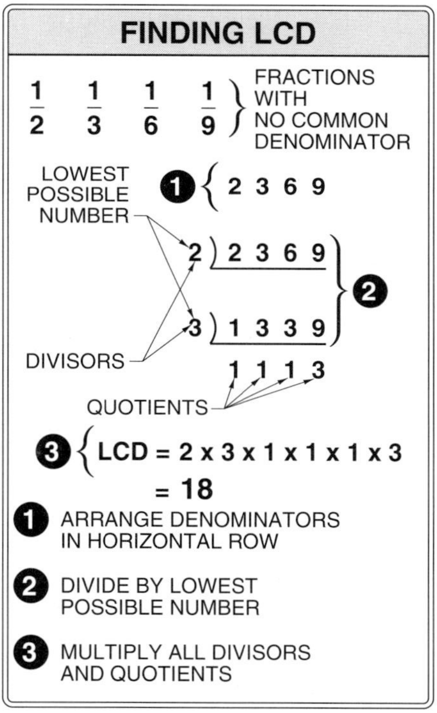

Figure 3-12. The lowest common denominator (LCD) is the smallest number into which the denominators of a group of two or more fractions divide an exact number of times.

For example, to find the LCD of $1/8$, $1/24$, and $1/9$, divide 8 and 24 by 2 ($8 \div 2 = 4$ and $24 \div 2 = 12$). Bring down the 9 and divide 4 and 12 by 2 ($4 \div 2 = 2$ and $12 \div 2 = 6$). Bring down the 9 and divide the 2 and 6 by 2 ($2 \div 2 = 1$ and $6 \div 2 = 3$). Bring down the 9 and divide the 3 and 9 by 3 (3 ÷ 3 = 1 and 9 ÷ 3 = 3). No further division is possible. Multiply the divisors and the quotients (2 × 2 × 2 × 3 × 1 × 1 × 3 = 72).

To check the LCD, divide it by all denominators of the original fractions. Each of these denominators should divide into the LCD an even number of times.

EXAMPLES

Finding LCD

1. Find the LCD for $\frac{1}{6}$, $\frac{3}{8}$, $\frac{2}{9}$, $\frac{5}{12}$, $\frac{5}{18}$, $\frac{7}{24}$, and $\frac{1}{36}$.

FRACTIONS
WITH NO COMMON
DENOMINATOR

$$\frac{1}{6} \ \frac{3}{8} \ \frac{2}{9} \ \frac{5}{12} \ \frac{5}{18} \ \frac{7}{24} \ \frac{1}{36}$$

❶ { 6 8 9 12 18 24 36

❷ {
2) 6 8 9 12 18 24 36
2) 3 4 9 6 9 12 18
2) 3 2 9 3 9 6 9
3) 3 1 9 3 9 3 9
3) 1 1 3 1 3 1 3
1 1 1 1 1 1 1

❸ { LCD = 2 x 2 x 2 x 3 x 3 x 1
= x 1 x 1 x 1 x 1 x 1 x 1
= **72**

❶ ARRANGE DENOMINATORS IN HORIZONTAL ROW
❷ DIVIDE BY LOWEST NUMBER POSSIBLE
❸ MULTIPLY ALL DIVISORS AND QUOTIENTS

① Arrange the denominators in a horizontal row. A division sign is drawn under the row.

② Divide 6, 8, 12, 18, 24, and 36 by 2 (6 ÷ 2 = 3, 8 ÷ 2 = 4, 12 ÷ 2 = 6, 18 ÷ 2 = 9, 24 ÷ 2 = 12, and 36 ÷ 2 = 18). Bring down the 9 because it cannot be divided by 2 an exact number of times.

② Divide 4, 6, 12, and 18 by 2 (4 ÷ 2 = 2, 6 ÷ 2 = 3, 12 ÷ 2 = 6, and 18 ÷ 2 = 9). Bring down the 3 and the 9s. ② Divide 2 and 6 by 2 (2 ÷ 2 = 1 and 6 ÷ 2 = 3). Bring down the 3 and the 9s. ② Divide the 3s and the 9s by 3. ② Divide the remaining 3s by 3.

③ Multiply the divisors and the quotients (2 × 2 × 2 × 3 × 3 × 1 × 1 × 1 × 1 × 1 × 1 × 1 = **72**).

2. Find the LCD of $\frac{2}{5}$, $\frac{1}{7}$, $\frac{4}{11}$, $\frac{6}{13}$, and $\frac{2}{3}$.

FRACTIONS WITH
PRIME NUMBERS
AS DENOMINATOR

$$\frac{2}{5} \ \frac{1}{7} \ \frac{4}{11} \ \frac{6}{13} \ \frac{2}{3}$$

❶ { 5 7 11 13 3

❷ { **NO DIVISION REQUIRED**

❸ { 5 x 7 x 11 x 13 x 3 = **15,015**

❶ ARRANGE DENOMINATORS IN HORIZONTAL ROW
❷ NO DIVISION REQUIRED
❸ MULTIPLY DENOMINATORS

All denominators in this example are prime numbers. No number divides into two or more of the denominators an exact number of times. To find the LCD for a group of different prime number denominators, all denominators are multiplied.

① Arrange the denominators in a horizontal row. ② Step 2 is not required because all denominators are prime numbers. ③ Multiply the denominators ($5 \times 7 \times 11 \times 13 \times 3 = $ **15,015**).

3. Find the LCD of $\frac{7}{36}$, $\frac{3}{25}$, $\frac{4}{39}$, $\frac{5}{27}$, and $\frac{9}{44}$.

FRACTIONS
WITH NO COMMON
DENOMINATOR

7	3	4	5	9
36	25	39	27	44

❶ { 36 25 39 27 44

❷ {
2) 36 25 39 27 44
2) 18 25 39 27 22
3) 9 25 39 27 11
3) 3 25 13   9 11
       1 25 13   3 11

❸ { LCD = 2 x 2 x 3 x 3 x 1 x 25
      = x 13 x 3 x 11
      = **386,100**

❶ ARRANGE DENOMINATORS IN HORIZONTAL ROW

❷ DIVIDE BY LOWEST NUMBER POSSIBLE

❸ MULTIPLY ALL DIVISORS AND QUOTIENTS

In this example 2 and 3 are used as divisors twice. No further division is possible because no number divides exactly into 1, 25, 13, 3, and 11.

① Arrange the denominators in a horizontal row. A division sign is drawn under the row. ② Divide 36 and 44 by 2 ($36 \div 2 = 18$ and $44 \div 2 = 22$). ② Divide 18 and 22 by 2 ($18 \div 2 = 9$ and $22 \div 2 = 11$). ② Divide 9 and 27 by 3 ($9 \div 3 = 3$ and $27 \div 3 = 9$). ② Divide 3 and 9 by 3 ($3 \div 3 = 1$ and $9 \div 3 = 3$). ③ Multiply the divisors and the quotients ($2 \times 2 \times 3 \times 3 \times 1 \times 25 \times 13 \times 3 \times 11 = $ **386,100**).

PRACTICE PROBLEMS

Finding LCD

Find the LCD.

1. $\frac{1}{6}$, $\frac{1}{8}$, $\frac{1}{12}$

2. $\frac{1}{12}$, $\frac{1}{16}$, $\frac{1}{24}$

3. $\frac{3}{10}$, $\frac{4}{15}$, $\frac{7}{20}$

4. $\frac{3}{7}$, $\frac{4}{11}$, $\frac{6}{7}$, $\frac{2}{13}$

5. $\frac{2}{3}$, $\frac{4}{5}$, $\frac{5}{11}$, $\frac{1}{5}$, $\frac{2}{17}$

6. $\frac{1}{2}$, $\frac{2}{3}$, $\frac{5}{16}$, $\frac{6}{20}$, $\frac{5}{15}$

7. $\frac{6}{30}$, $\frac{15}{18}$, $\frac{10}{55}$, $\frac{16}{56}$

8. $\frac{3}{5}$, $\frac{4}{15}$, $\frac{10}{20}$, $\frac{6}{25}$

9. $\frac{1}{12}$, $\frac{3}{6}$, $\frac{4}{15}$, $\frac{1}{10}$

10. $\frac{1}{12}$, $\frac{1}{8}$, $\frac{1}{6}$

Reducing Fractions to LCD. The purpose of reducing a group of fractions to LCD is to give all of the fractions a common denominator. This procedure is exactly the same and produces the same result as changing to higher terms with a given denominator.

The given denominator is the LCD. Changing or reducing a fraction to higher or lower terms is performed without changing the value of the fraction.

To reduce fractions to the LCD, divide the LCD by each denominator and multiply each term of the fraction (both numerator and denominator) by the quotient. For example, to reduce the fractions $\frac{3}{5}$ and $\frac{5}{6}$ to their LCD, divide the LCD (30) by each denominator (30 ÷ 5 = 6 and 30 ÷ 6 = 5).

Multiply the terms of the fractions by the quotients ($\frac{3 \times 6}{5 \times 6} = {}^{18}\!/_{30}$ and $\frac{5 \times 5}{6 \times 5} = {}^{25}\!/_{30}$). To check fractions reduced to the LCD, the fractions are reduced to their lowest terms.

PRACTICE PROBLEMS

Reducing Fractions to LCD

Reduce to the LCD.
1. $\frac{1}{7}$, $\frac{3}{8}$, $\frac{1}{4}$, $\frac{2}{3}$
2. $\frac{2}{7}$, $\frac{1}{3}$, $\frac{3}{4}$, $\frac{1}{2}$
3. $\frac{1}{3}$, $\frac{2}{9}$, $\frac{5}{12}$, $\frac{3}{8}$
4. $\frac{1}{15}$, $\frac{3}{10}$, $\frac{4}{25}$, $\frac{1}{30}$
5. $\frac{1}{3}$, $\frac{2}{7}$, $\frac{4}{11}$

ADDING FRACTIONS

Only like objects can be added to make one sum. It is impossible to add apples to oranges, cents to dollars, screws to bolts, etc. For example, a person has three dollar bills and 400 pennies. The dollar bills are simply counted, but dollars cannot be added to pennies because they are entirely different in name and value per unit.

To add dissimilar quantities, one of the quantities is changed to terms of the other. For example, to add pennies to dollars, the pennies are changed into dollars, and then the dollars are added.

One dollar contains 100 pennies, so the 400 pennies are divided by 100 (400 ÷ 100 = $4.00). Add the $4.00 to the $3.00 ($4.00 + $3.00 = **$7.00**).

Another example of changing dissimilar objects to common terms is trying to add 20 hex head machine screws to 35 carriage bolts. Adding screws and bolts cannot be done because they are different objects. They must be changed to a common term.

The term "fasteners" describes both machine screws and carriage bolts. The 20 machine screws can be referred to as 20 fasteners. The 35 carriage bolts can be referred to as 35 fasteners. By changing the screws and bolts to the common term "fasteners," they can be added to find the sum (20 + 35 = **55 fasteners**).

This technique is also used for the addition of fractions. For example, there are two $\frac{1}{2}$ gallons of paint. These halves added together yield a whole gallon because $\frac{1}{2} + \frac{1}{2} = 1$.

However, if $\frac{1}{4}$ gallon of paint and $\frac{1}{3}$ gallon of paint are combined, the fractions cannot be added as $\frac{1}{4} + \frac{1}{3}$ even though they are expressed with the common term "gallons."

The denominator of $\frac{1}{4}$ indicates the gallon is divided into four parts, and the denominator of $\frac{1}{3}$ indicates the gallon is divided into three parts. Before these fractions can be added, they must have the same denominator. In any group of fractions being added, the denominators must indicate that all objects have been divided into the same number of parts.

Adding Like Denominators

To add any group of fractions with a common denominator, add the numerators and place the sum over the denominator. See Figure 3-13. For example, to add $\frac{1}{5} + \frac{4}{5} + \frac{3}{5}$ add the numerators (1 + 4 + 3 = 8). Place the 8 over 5 ($\frac{8}{5} = \mathbf{1\frac{3}{5}}$). The fraction $\frac{8}{5}$ is an improper fraction and reduces to the mixed number $1\frac{3}{5}$.

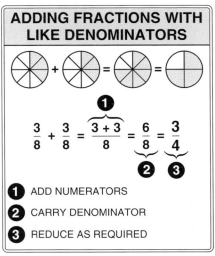

ADDING FRACTIONS WITH LIKE DENOMINATORS

❶

$$\frac{3}{8} + \frac{3}{8} = \frac{3+3}{8} = \frac{6}{8} = \frac{3}{4}$$

❷ ❸

❶ ADD NUMERATORS

❷ CARRY DENOMINATOR

❸ REDUCE AS REQUIRED

Figure 3-13. To add any group of fractions with a common denominator, add the numerators and place the sum over the denominator.

EXAMPLES

Adding Like Denominators

1. Add $\frac{1}{4} + \frac{2}{4} + \frac{3}{4}$.

$$\frac{1}{4} + \frac{2}{4} + \frac{3}{4}$$

$$= \frac{\overset{❶}{1+2+3}}{4} = \frac{\overset{❷}{6}}{4} = 1\frac{2}{4} = 1\frac{\overset{❸}{1}}{2}$$

❶ ADD NUMERATORS
❷ CARRY DENOMINATOR
❸ REDUCE AS REQUIRED

① Add the numerators (1 + 2 + 3 = 6). ② Place the 6 over 4 ($\frac{6}{4}$). ③ The fraction $\frac{6}{4}$ is an improper fraction and reduces to the mixed number $1\frac{2}{4}$. The fraction $1\frac{2}{4}$ reduces to $\frac{1}{2}$ ($1\frac{6}{4} = 1\frac{2}{4} = \mathbf{1\frac{1}{2}}$).

2. Add $\frac{1}{16} + \frac{3}{16} + \frac{9}{16} + \frac{7}{16} + \frac{11}{16}$.

$$\frac{1}{16} + \frac{3}{16} + \frac{9}{16} + \frac{7}{16} + \frac{11}{16}$$

$$= \frac{\overbrace{1 + 3 + 9 + 7 + 11}^{\textbf{1}}}{16}$$

$$= \frac{\overbrace{31}^{\textbf{2}}}{16} = 1\frac{\overbrace{15}^{\textbf{3}}}{16}$$

❶ ADD NUMERATORS
❷ CARRY DENOMINATOR
❸ REDUCE AS REQUIRED

① Add the numerators (1 + 3 + 9 + 7 + 11 = 31). ② Place the 31 over 16 ($\frac{31}{16}$). ③ Reduce the improper fraction ($\frac{31}{16} = \mathbf{1\frac{15}{16}}$).

PRACTICE PROBLEMS

Adding Like Denominators

1. $\frac{1}{5} + \frac{2}{5} + \frac{3}{5}$
2. $\frac{8}{9} + \frac{4}{9} + \frac{7}{9} + \frac{2}{9}$
3. $\frac{1}{24} + \frac{5}{24} + \frac{19}{24} + \frac{13}{24}$
4. A painter has four 1 gallon cans of paint. The cans are $\frac{1}{8}$, $\frac{3}{8}$, $\frac{5}{8}$ and $\frac{7}{8}$ full. How much paint does the painter have?

5. Find the center-to-center distance between the drilled holes of Bracket A.

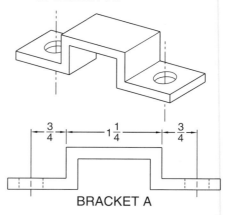

BRACKET A

Adding Unlike Denominators

Fractions with unlike denominators must be changed to have a common denominator before they can be added. The LCD is found in order to add fractions with unlike denominators. See Figure 3-14.

For example, to add $\frac{3}{4} + \frac{5}{6} + \frac{7}{8} + \frac{5}{12}$, the fractions are reduced to common terms (LCD) or changed to have common denominators. The fractions have an LCD of 24. The reduced fractions are $\frac{3}{4} = \frac{18}{24}$, $\frac{5}{6} = \frac{20}{24}$, $\frac{7}{8} = \frac{21}{24}$, and $\frac{5}{12} = \frac{10}{24}$. Add the numerators (18 + 20 + 21 + 10 = 69). Place the 69 over 24 ($\frac{69}{24} = 2\frac{21}{24} = \mathbf{2\frac{7}{8}}$). The fraction $\frac{69}{24}$ is an improper fraction and reduces to the mixed number $2\frac{21}{24}$. The fraction $\frac{21}{24}$ reduces to $\frac{7}{8}$.

ADDING FRACTIONS WITH UNLIKE DENOMINATORS

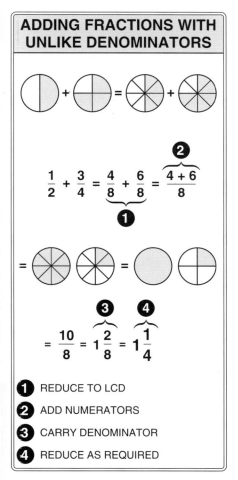

$$\frac{1}{2} + \frac{3}{4} = \frac{4}{8} + \frac{6}{8} = \overbrace{\frac{4+6}{8}}^{\textbf{2}}$$

①

$$= \frac{10}{8} = 1\overbrace{\frac{2}{8}}^{\textbf{3}} = 1\overbrace{\frac{1}{4}}^{\textbf{4}}$$

❶ REDUCE TO LCD

❷ ADD NUMERATORS

❸ CARRY DENOMINATOR

❹ REDUCE AS REQUIRED

Figure 3-14. The LCD must be found before adding fractions with unlike denominators.

EXAMPLES

Adding Unlike Denominators

1. Add $\frac{2}{3} + \frac{7}{18} + \frac{20}{26} + \frac{27}{21}$.

$$\left.\frac{2}{3} + \frac{7}{18} + \frac{20}{26} + \frac{27}{21}\right\} \begin{array}{l}\text{FRACTIONS} \\ \text{WITH UNLIKE} \\ \text{DENOMINATORS}\end{array}$$

$$\textbf{❶}\begin{cases}\dfrac{2}{3} = \dfrac{1092}{1638} & \dfrac{7}{18} = \dfrac{637}{1638} \\[2ex] \dfrac{20}{26} = \dfrac{1260}{1638} & \dfrac{27}{21} = \dfrac{2106}{1638}\end{cases} \overset{\text{LCD}}{}$$

$$\textbf{❷}\left\{\dfrac{1092 + 637 + 1260 + 2106}{1638}\right.$$

$$= \underbrace{\dfrac{5095}{1638}}_{\textbf{❸}} = 3\overbrace{\dfrac{181}{1638}}^{\textbf{❹}}$$

❶ REDUCE TO LCD
❷ ADD NUMERATORS
❸ CARRY DENOMINATOR
❹ REDUCE AS REQUIRED

① Reduce fractions to the LCD, which is 1638. The reduced fractions are $\frac{2}{3} = \frac{1092}{1638}$, $\frac{7}{18} = \frac{637}{1638}$, $\frac{20}{26} = \frac{1260}{1638}$, and $\frac{27}{21} = \frac{2106}{1638}$. ② Add the numerators $(1092 + 637 + 1260 + 2106 = 5095)$. ③ Place the 5095 over 1638 ($\frac{5095}{1638}$). ④ Reduce the improper fraction ($\frac{5095}{1638} = 3\frac{181}{1638}$). The fraction $\frac{5095}{1638}$ reduces to the mixed number $3\frac{181}{1638}$.

To simplify the problem, the fractions can be reduced to lowest terms before finding the LCD. For example, the fractions $\frac{20}{26}$ and $\frac{27}{21}$ reduce to $\frac{10}{13}$ and $\frac{9}{7}$. The problem is now simplified as $\frac{2}{3} + \frac{7}{18} + \frac{10}{13} + \frac{9}{7}$.

2. Add $\frac{7}{8} + \frac{2}{5} + \frac{3}{10}$.

$$\left.\frac{7}{8} + \frac{2}{5} + \frac{3}{10} \right\} \begin{array}{l}\text{FRACTIONS}\\ \text{WITH UNLIKE}\\ \text{DENOMINATORS}\end{array}$$

❶ $\begin{cases} \dfrac{7}{8} = \dfrac{X}{40} \;\text{—LCD} & \dfrac{2}{5} = \dfrac{X}{40} & \dfrac{3}{10} = \dfrac{X}{40} \\[2mm] 40 \div 8 = 5 & 40 \div 5 = 8 & 40 \div 10 = 4 \\[2mm] \dfrac{7}{5} = \dfrac{7 \times 5}{8 \times 5} = \dfrac{35}{40} & \dfrac{2}{5} = \dfrac{2 \times 8}{5 \times 8} = \dfrac{16}{40} & \dfrac{3}{10} = \dfrac{3 \times 4}{10 \times 4} = \dfrac{12}{40} \end{cases}$

$$\overbrace{\frac{35 + 16 + 12}{40}}^{\textbf{❷}} = \underbrace{\frac{63}{40}}_{\textbf{❸}} = 1\overbrace{\frac{23}{40}}^{\textbf{❹}}$$

❶ REDUCE TO LCD   **❸** CARRY DENOMINATOR
❷ ADD NUMERATORS   **❹** REDUCE AS REQUIRED

① Reduce fractions to the LCD, which is 40. The reduced fractions are $\frac{7}{8} = \frac{35}{40}$, $\frac{2}{5} = \frac{16}{40}$, and $\frac{3}{10} = \frac{12}{40}$. ② Add the numerators ($35 + 16 + 12 = 63$). ③ Place the 63 over 40 ($\frac{63}{40}$). ④ Reduce the improper fraction ($\frac{63}{40} = 1\frac{23}{40}$). The fraction $\frac{63}{40}$ reduces to the mixed number $1\frac{23}{40}$.

PRACTICE PROBLEMS

Adding Unlike Denominators

1. $\frac{1}{5} + \frac{1}{4} + \frac{1}{6}$
2. $\frac{1}{2} + \frac{5}{8} + \frac{9}{10}$
3. $\frac{3}{2} + \frac{7}{9} + \frac{2}{3}$
4. $\frac{31}{24} + \frac{11}{12} + \frac{3}{8}$
5. $\frac{25}{24} + \frac{5}{6} + \frac{7}{8}$

Adding Mixed Numbers

A *mixed number* is a combination of a whole number and a proper fraction. To add mixed numbers, add the whole numbers, add the numerators, carry the denominator, and reduce as required. See Figure 3-15. If the sum of the fractions is an improper fraction, reduce to a mixed number. Add the whole number of the mixed number to the sum of the whole numbers.

For example, to add mixed numbers with unlike denominators, such as $1\frac{1}{2}$, $2\frac{2}{5}$, and $3\frac{19}{20}$, add the whole numbers ($1 + 2 + 3 = 6$). To add the fractions, the LCD (20) is found. Add the reduced fractions of $\frac{1}{2} = \frac{10}{20}$, $\frac{2}{5} = \frac{8}{20}$, and $\frac{19}{20}$ ($\frac{10}{20} + \frac{8}{20} + \frac{19}{20} = \frac{37}{20}$). The fraction

$^{37}/_{20}$ is an improper fraction and is reduced to a mixed number ($^{37}/_{20}$ = $1^{17}/_{20}$). Add the sum of the whole numbers to the sum of the fractions ($6 + 1^{17}/_{20} = 7^{17}/_{20}$).

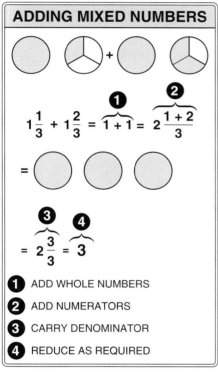

ADDING MIXED NUMBERS

❶ ADD WHOLE NUMBERS
❷ ADD NUMERATORS
❸ CARRY DENOMINATOR
❹ REDUCE AS REQUIRED

Figure 3-15. To add mixed numbers, add the whole numbers, add the numerators, carry the denominator, and reduce as required.

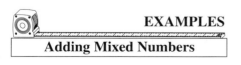

EXAMPLES

Adding Mixed Numbers

1. Add $3\frac{1}{3} + 2\frac{1}{2} + 6\frac{1}{6}$.

$$3\frac{1}{3} + 2\frac{1}{2} + 6\frac{1}{6} \left.\right\} \begin{matrix} \text{MIXED NUMBERS} \\ \text{WITH UNLIKE} \\ \text{DENOMINATORS} \end{matrix}$$

❶ $\left\{ 3 + 2 + 6 = 11 \right.$

❷ $\left\{ \begin{array}{l} 3 \overline{)3\ 2\ 6} \\ 2 \overline{)1\ 2\ 2} \\ \quad\ 1\ 1\ 1 \\ 3 \times 2 \times 1 \times 1 \times 1 = 6 \\ \text{LCD = 6} \\ \frac{1}{3} = \frac{2}{6} \quad \frac{1}{2} = \frac{3}{6} \quad \frac{1}{6} = \frac{1}{6} \end{array} \right.$

$$\overset{❸}{\overbrace{2 + 3 + 1}} = \frac{\overset{❹}{6}}{6} = \overset{❺}{1}$$

$$\overset{❻}{\overbrace{11 + 1}} = 12$$

❶ ADD WHOLE NUMBERS
❷ FIND LCD AND REDUCE FRACTIONS
❸ ADD NUMERATORS
❹ CARRY DENOMINATORS
❺ REDUCE AS REQUIRED
❻ ADD SUMS OF WHOLE NUMBERS AND FRACTIONS

① Add the whole numbers ($3 + 2 + 6 = 11$). ② Find the LCD (6) for the fractions. Reduce the fractions ($^1/_3 = ^2/_6$, $^1/_2 = ^3/_6$, and $^1/_6 = ^1/_6$). ③ Add the numerators ($2 + 3 + 1 = 6$). ④ Place the 6 over 6. ⑤ Reduce as required ($^6/_6 = 1$). ⑥ Add the sum of the whole numbers to the sum of the fractions ($11 + 1 = $ **12**).

2. Add $1\frac{1}{4} + 2\frac{3}{4}$.

$$1\frac{1}{4} + 2\frac{3}{4} \left.\right\} \begin{array}{l}\text{MIXED} \\ \text{NUMBERS} \\ \text{WITH LIKE} \\ \text{DENOMINATORS}\end{array}$$

$$1\frac{1}{4} + 2\frac{3}{4} = \overbrace{1 + 2}^{\text{❶}} = 3\overbrace{\frac{1 + 3}{4}}^{\text{❷}}$$

$$= \overbrace{3\frac{4}{4}}^{\text{❸}} = \overbrace{4}^{\text{❹}}$$

❶ ADD WHOLE NUMBERS
❷ ADD NUMERATORS
❸ CARRY DENOMINATORS
❹ REDUCE AS REQUIRED

 ① Add the whole numbers (1 + 2). ② Add the numerators (1 + 3). ③ Place the 4 over 4. ④ Reduce as required (3⁴⁄₄ = **4**).

PRACTICE PROBLEMS

Adding Mixed Numbers

1. $2\frac{5}{8} + 3\frac{7}{12} + 5\frac{11}{24}$
2. $1\frac{3}{20} + 2\frac{7}{12} + 3\frac{5}{15}$
3. $6\frac{2}{3} + 2\frac{2}{7} + 4\frac{4}{15}$
4. $3\frac{5}{12} + \frac{7}{12} + 2\frac{9}{24}$
5. $5\frac{1}{2} + \frac{3}{4} + 6\frac{2}{3}$
6. $9\frac{2}{3} + 7\frac{7}{8} + 12\frac{3}{4}$
7. $8\frac{4}{5} + \frac{6}{7} + 4\frac{5}{8}$
8. $6\frac{2}{7} + 2\frac{1}{3} + 5\frac{3}{4}$
9. Find the total length of the five pieces of pipe on the Plumbing Riser.

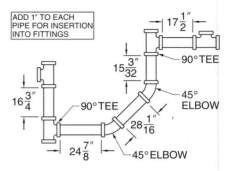

PLUMBING RISER

10. Find the total length of four table legs from the Table Leg.

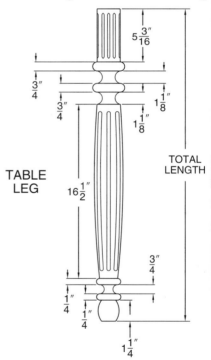

TABLE LEG

SUBTRACTING FRACTIONS

Before fractions can be subtracted from one another, they must have common denominators. Reduce all fractions to be subtracted to their LCD. When mixed numbers are subtracted, the whole numbers and the fractions are subtracted separately and the two differences are added.

Subtracting Like Denominators

To subtract fractions that have common denominators, subtract the numerators and place the difference over the denominator. See Figure 3-16. For example, to subtract $\frac{5}{16}$ from $\frac{9}{16}$, subtract 5 from 9 ($9 - 5 = 4$). Place the 4 over 16 ($\frac{4}{16}$). Reduce $\frac{4}{16}$ by dividing the numerator and denominator by 4 ($\frac{4 \div 4}{16 \div 4} = \frac{1}{4}$).

SUBTRACTING FRACTIONS WITH LIKE DENOMINATORS

$$\frac{7}{8} - \frac{5}{8} = \overbrace{\frac{7-5}{8}}^{\text{❶}} = \underbrace{\frac{2}{8}}_{\text{❷}} = \underbrace{\frac{1}{4}}_{\text{❸}}$$

❶ SUBTRACT NUMERATORS

❷ CARRY DENOMINATOR

❸ REDUCE AS REQUIRED

Figure 3-16. To subtract fractions that have common denominators, subtract the numerators and place the difference over the denominator.

EXAMPLES

Subtracting Like Denominators

1. Subtract $\frac{2}{8}$ from $\frac{7}{8}$.

$$\frac{7}{8} - \frac{2}{8} = \overbrace{\frac{7-2}{8}}^{\text{❶}} = \underbrace{\frac{5}{8}}_{\text{❷}}$$

❶ SUBTRACT NUMERATORS
❷ CARRY DENOMINATOR

① Subtract 2 from 7 ($7 - 2 = 5$).
② Place the 5 over 8 ($\frac{7}{8} - \frac{2}{8} = \frac{5}{8}$).

2. Subtract $\frac{5}{24}$ from $\frac{23}{24}$.

$$\frac{23}{24} - \frac{5}{24} = \overbrace{\frac{23-5}{24}}^{\text{❶}} = \underbrace{\frac{18}{24}}_{\text{❷}} = \underbrace{\frac{3}{4}}_{\text{❸}}$$

❶ SUBTRACT NUMERATORS
❷ CARRY DENOMINATOR
❸ REDUCE AS REQUIRED

① Subtract 5 from 23 ($23 - 5 = 18$). ② Place the 18 over 24 ($\frac{18}{24}$). ③ Reduce $\frac{18}{24}$ by dividing the numerator and denominator by 6 ($\frac{18 \div 6}{24 \div 6} = \frac{3}{4}$).

3. Subtract $\frac{5}{45}$ from $\frac{41}{45}$.

$$\frac{41}{45} - \frac{5}{45} = \overbrace{\frac{41-5}{45}}^{\text{❶}} = \underbrace{\frac{36}{45}}_{\text{❷}} = \underbrace{\frac{4}{5}}_{\text{❸}}$$

❶ SUBTRACT NUMERATORS
❷ CARRY DENOMINATOR
❸ REDUCE AS REQUIRED

① Subtract 5 from 41 (41 − 5 = 36). ② Place the 36 over 45 ($^{36}/_{45}$). ③ Reduce $^{36}/_{45}$ by dividing the numerator and denominator by 9 ($\frac{36 \div 9}{45 \div 9}$ = $^4/_5$).

PRACTICE PROBLEMS

Subtracting Like Denominators

1. $^8/_9 - ^5/_9$
2. $^{14}/_{12} - ^{11}/_{12}$
3. $^{20}/_{27} - ^6/_{27}$
4. $^{49}/_{70} - ^{36}/_{70}$
5. Find the missing dimension (X) on the Shaft Gauge.

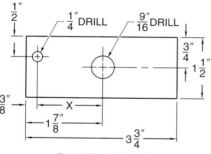

SHAFT GAUGE

Subtracting Unlike Denominators

To subtract fractions having unlike denominators, reduce the fractions to the LCD. Subtract one numerator from the other. See Figure 3-17. For example, to subtract $^1/_6$ from $^7/_{12}$, reduce the fractions to the LCD 12 ($^1/_6 = ^2/_{12}$ and $^7/_{12} = ^7/_{12}$). Subtract the numerators (7 − 2 = 5). Place the 5 over 12 ($^7/_{12} - ^1/_6 = ^5/_{12}$).

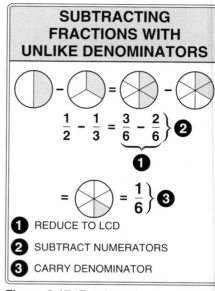

SUBTRACTING FRACTIONS WITH UNLIKE DENOMINATORS

$\frac{1}{2} - \frac{1}{3} = \frac{3}{6} - \frac{2}{6}$ } ②

①

$= \frac{1}{6}$ } ③

❶ REDUCE TO LCD
❷ SUBTRACT NUMERATORS
❸ CARRY DENOMINATOR

Figure 3-17. To subtract fractions that have unlike denominators, reduce to the LCD and subtract one numerator from the other.

EXAMPLES

Subtracting Unlike Denominators

1. Subtract $^7/_{12}$ from $^{23}/_{36}$.

$\frac{23}{36} - \frac{7}{12}$ } FRACTIONS WITH UNLIKE DENOMINATORS

LCD

❶ { $\frac{7}{12} = \frac{21}{36}$ / $\frac{23}{36} = \frac{23}{36}$

❷
$\frac{23 - 21}{36} = \frac{2}{36} = \frac{1}{18}$ ❹

❸

❶ REDUCE TO LCD
❷ SUBTRACT NUMERATORS
❸ CARRY DENOMINATOR
❹ REDUCE AS REQUIRED

① Reduce the fractions to the LCD ($\frac{7}{12} = \frac{21}{36}$ and $\frac{23}{36} = \frac{23}{36}$). ② Subtract 21 from 23 (23 − 21 = 2). ③ Place the 2 over 36 ($\frac{2}{36}$). ④ Reduce $\frac{2}{36}$ by dividing the numerator and denominator by 2 ($\frac{2 \div 2}{36 \div 2} = \frac{1}{18}$).

2. Subtract $\frac{7}{11}$ from $\frac{29}{43}$.

Note: The denominators in this example are prime numbers. To find the LCD when the denominators are prime numbers, multiply the denominators.

$$\frac{29}{43} - \frac{7}{11} \left\} \begin{array}{l}\text{FRACTIONS}\\\text{WITH PRIME}\\\text{NUMBERS AS}\\\text{DENOMINATORS}\end{array}\right.$$

$$❶\begin{cases}\text{LCD} = 43 \times 11 = 473\\\frac{7}{11} = \frac{301}{473} \quad \frac{29}{43} = \frac{319}{473}\end{cases}$$

$$❷\begin{cases}\frac{319 - 301}{473} = \frac{18}{473}\end{cases}❸$$

❶ REDUCE TO LCD
❷ SUBTRACT NUMERATORS
❸ CARRY DENOMINATOR

① Multiply 11 by 43 to find the LCD (11 × 43 = 473). Reduce fractions to the LCD ($\frac{7}{11} = \frac{301}{473}$ and $\frac{29}{43} = \frac{319}{473}$). ② Subtract the 301 from 319 (319 − 301 = 18). ③ Place the 18 over 473 ($\frac{18}{473}$). The fraction $\frac{18}{473}$ is in the lowest terms.

3. Subtract $\frac{3}{5}$ from $\frac{7}{8}$.

Note: No number divides into both denominators. To find the LCD, multiply the denominators.

$$\frac{7}{8} - \frac{3}{5} = ❶\begin{cases}\text{LCD} = 8 \times 5 = 40\\\frac{3}{5} = \frac{24}{40} \quad \frac{7}{8} = \frac{35}{40}\end{cases}$$

$$❷\begin{cases}\frac{35 - 24}{40} = \frac{11}{40}\end{cases}❸$$

❶ REDUCE TO LCD
❷ SUBTRACT NUMERATORS
❸ CARRY DENOMINATOR

① Multiply 8 by 5 to find the LCD (8 × 5 = 40). Reduce fractions to the LCD ($\frac{3}{5} = \frac{24}{40}$ and $\frac{7}{8} = \frac{35}{40}$). ② Subtract the 24 from 35 (35 − 24 = 11). ③ Place the 11 over 40 ($\frac{11}{40}$). The fraction $\frac{11}{40}$ is in the lowest terms.

4. Subtract $\frac{3}{4}$ from $\frac{24}{10}$.

Note: The $\frac{24}{10}$ can be reduced to lower terms before finding the LCD. The $\frac{24}{10}$ reduces to $\frac{12}{5}$. The problem becomes $\frac{12}{5} - \frac{3}{4}$.

$$\frac{24}{10} - \frac{3}{4} = \frac{12}{5} - \frac{3}{4}$$

$$❶\begin{cases}\text{LCD} = 4 \times 5 = 20\\\frac{3}{4} = \frac{15}{20} \quad \frac{12}{5} = \frac{48}{20}\end{cases}$$

$$❷\begin{cases}\frac{48 - 15}{20} = \frac{33}{20} = 1\frac{13}{20}\end{cases}❹$$
❸

❶ REDUCE TO LCD
❷ SUBTRACT NUMERATORS
❸ CARRY DENOMINATOR
❹ REDUCE AS REQUIRED

① Multiply 4 by 5 to find the LCD ($4 \times 5 = 20$). Reduce fractions to the LCD ($\frac{3}{4} = \frac{15}{20}$ and $\frac{24}{10} = \frac{48}{20}$). ② Subtract the 15 from 48 ($48 - 15 = 33$). ③ Place the 33 over 20 ($\frac{33}{20}$). ④ Reduce the improper fraction to a mixed number ($\frac{33}{20} = \mathbf{1\frac{13}{20}}$).

PRACTICE PROBLEMS

Subtracting Unlike Denominators

1. Subtract $\frac{4}{25}$ from $\frac{3}{8}$.
2. Subtract $\frac{4}{35}$ from $\frac{84}{120}$.
3. Subtract $\frac{4}{5}$ from $\frac{6}{7}$.
4. Subtract $\frac{3}{8}$ from $\frac{17}{32}$.

5. Find the final thickness of Board A.

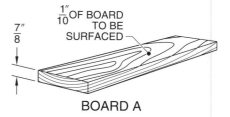

BOARD A

Subtracting Mixed Numbers

To subtract mixed numbers, subtract the whole numbers and fractions separately. Add the difference of the whole numbers and the difference of the fractions to find the answer.

If the fractions of the mixed numbers do not have a common denominator, they must be reduced to the LCD before subtraction. See Figure 3-18.

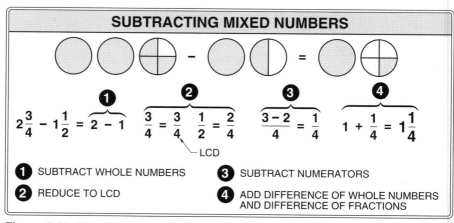

Figure 3-18. To subtract mixed numbers, subtract the whole numbers and the fractions separately.

For example, to subtract $1\frac{1}{16}$ from $5\frac{3}{16}$, subtract the whole numbers ($5 - 1 = 4$). Subtract $\frac{1}{16}$ from $\frac{3}{16}$ ($\frac{3}{16} - \frac{1}{16} = \frac{2}{16}$).

Add the difference of the whole numbers and the difference of the fractions ($4 + \frac{2}{16} = 4\frac{2}{16} = \mathbf{4\frac{1}{8}}$). The fraction $\frac{2}{16}$ is not in lowest terms and reduces to $\frac{1}{8}$.

EXAMPLE

Subtracting Mixed Numbers

1. Subtract $4\frac{1}{2}$ from $10\frac{2}{3}$.

$$10\frac{2}{3} - 4\frac{1}{2}$$

❶ $\{10 - 4 = 6$

❷ $\left\{\dfrac{1}{2} = \dfrac{3}{6} \quad \dfrac{2}{3} = \dfrac{4}{6}\right.$ — LCD

❸ $\left\{\dfrac{4 - 3}{6} = \dfrac{1}{6}\right.$

❹ $\left\{6 + \dfrac{1}{6} = 6\dfrac{1}{6}\right.$

❶ SUBTRACT WHOLE NUMBERS
❷ REDUCE TO LCD
❸ SUBTRACT NUMERATORS
❹ ADD DIFFERENCE OF WHOLE NUMBERS AND DIFFERENCE OF FRACTIONS

① Subtract 4 from 10 ($10 - 4 = 6$). ② Reduce the fractions to the LCD ($\frac{1}{2} = \frac{3}{6}$ and $\frac{2}{3} = \frac{4}{6}$). ③ Subtract the $\frac{3}{6}$ from $\frac{4}{6}$ ($\frac{4}{6} - \frac{3}{6} = \frac{1}{6}$). ④ Add the difference of the whole

numbers and the difference of the fractions ($6 + \frac{1}{6} = \mathbf{6\frac{1}{6}}$).

PRACTICE PROBLEMS

Subtracting Mixed Numbers

1. Subtract $2\frac{1}{2}$ from $5\frac{3}{4}$.
2. Subtract $8\frac{2}{9}$ from $14\frac{5}{18}$.
3. Subtract $3\frac{3}{5}$ from $6\frac{7}{8}$.
4. Subtract $1\frac{3}{10}$ from $4\frac{2}{3}$.
5. A carpenter measures and saws a piece of Baseboard Molding into three pieces. Disregarding the kerf (saw waste), find the length of the third piece (D).

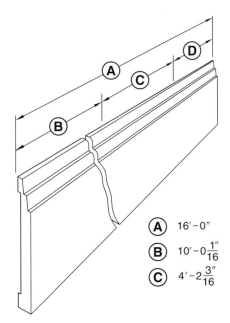

(A) $16'-0''$

(B) $10'-0\frac{1}{16}''$

(C) $4'-2\frac{3}{16}''$

BASEBOARD MOLDING

6. A service station has a 1000 gal. tank containing $949\frac{7}{8}$ gal. of gasoline. The service station sold 170⅔ gal., 215⅝ gal., 167⅞ gal., and 223⅕ gal. on four consecutive days. How many gallons are needed to fill the tank? *(Round to the next lower full number)*

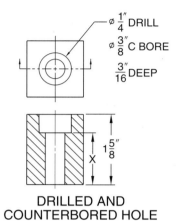

DRILLED AND
COUNTERBORED HOLE

7. A machinist has a steel bar 5¼′ long from which pieces 2½′ long and 2⅙′ long are cut. Disregarding the saw kerf, find the length of the remaining steel bar.

8. A contractor has 24½ bags of cement mix for a job. Only 17¼ bags are needed. How many bags of cement mix are left?

9. Find the missing dimension X on the Drilled and Counterbored Hole.

10. In estimating labor cost for a job, a mason figured a total of 42 hours. The job was underestimated. The hours worked were 7¾, 8⅙, 9⅔, 10½, 9½, and 13⅚. Find by how many hours the job was underestimated.

Name _____ Date _____

True-False and Completion

T   F   **1.** An improper fraction is a fraction that has a numerator larger than the denominator.

_____   **2.** The _____ shows into how many parts a whole number is divided.

T   F   **3.** Only fractions with like denominators can be added.

_____   **4.** A(n) _____ is one part of a whole number.

T   F   **5.** The whole numbers and the fractions are subtracted separately when subtracting mixed numbers.

T   F   **6.** The 3 is the denominator in the fraction $\frac{3}{4}$.

_____   **7.** A(n) _____ is a combination of a whole number and a proper fraction.

T   F   **8.** The fraction $1\frac{5}{9}$ is an improper fraction.

_____   **9.** The _____ of the fraction are the numerator and denominator.

_____   **10.** The _____ is the smallest number into which the denominators of a group of two or more fractions divide an exact number of times.

Calculations

Change the following fractions by multiplying the terms by 3.

_____   **1.** $\frac{2}{3}$

_____   **2.** $\frac{17}{32}$

_____   **3.** $^{25}/_{64}$

_____   **4.** Subtract $^{3}/_{32}$ from $^{9}/_{16}$.

_____   **5.** Add $^{3}/_{8}$ + $^{3}/_{8}$ + $^{3}/_{8}$.

_____   **6.** A machinist cuts five pieces from a 25″ steel bar. Find the amount of steel bar left if the five pieces are $3\frac{1}{2}$″, $3\frac{3}{4}$″, $4\frac{3}{4}$″, $4\frac{1}{4}$″, and $4\frac{2}{3}$″ long.

_____   **7.** Subtract $7^{7}/_{16}$ from $25^{23}/_{32}$.

_____   **8.** A board is $5^{9}/_{16}$″ wide. A piece $2^{3}/_{8}$″ is ripped off the board. The saw kerf is $^{1}/_{8}$″. What is the remaining width of the board?

_____   **9.** Add $3\frac{3}{4}$ and $2\frac{1}{8}$ and subtract the sum from $19^{3}/_{32}$.

_____   **10.** Add $^{3}/_{16}$ and $^{5}/_{32}$ and subtract the sum from $^{7}/_{8}$.

_____   **11.** Subtract $1^{7}/_{16}$ from $2^{1}/_{16}$ and add the difference to $4\frac{3}{4}$.

Change improper fractions to mixed numbers and mixed numbers to improper fractions.

_____   **12.** $^{60}/_{16}$

_____   **13.** $^{75}/_{32}$

_____   **14.** $6\frac{3}{4}$

_____   **15.** $5^{7}/_{16}$

Add the following.

_____   **16.** $^{3}/_{8}$ + $1^{7}/_{16}$ + $4\frac{3}{4}$

_____   **17.** $^{7}/_{32}$ + $^{9}/_{16}$ + $^{3}/_{8}$

_____   **18.** $12\frac{1}{2}$ + $2^{1}/_{16}$ + $5\frac{7}{8}$

Subtract the following.

_____ **19.** $^{19}\!/_{32} - {}^{3}\!/_{16}$

_____ **20.** $6^{3}\!/_{8} - 4^{15}\!/_{16}$

_____ **21.** $7^{3}\!/_{4} - {}^{5}\!/_{16}$

Reduce to the LCD.

_____ **22.** $^{7}\!/_{8}, {}^{3}\!/_{4}, {}^{2}\!/_{3}, {}^{5}\!/_{16}$

_____ **23.** $^{7}\!/_{8}, {}^{3}\!/_{4}, {}^{5}\!/_{16}, {}^{9}\!/_{32}$

_____ **24.** Add $^{3}\!/_{16} + {}^{19}\!/_{64} + {}^{21}\!/_{32} + {}^{7}\!/_{8}$.

_____ **25.** Subtract $10^{3}\!/_{4}$ from $13^{3}\!/_{5}$.

Reduce to lower terms.

_____ **26.** $^{24}\!/_{36}$

_____ **27.** $^{40}\!/_{64}$

Reduce to the lowest possible terms.

_____ **28.** $^{20}\!/_{32}$

_____ **29.** $^{14}\!/_{16}$

_____ **30.** What length copper pipe is required so that two $3^{3}\!/_{4}''$ pieces, four $8^{2}\!/_{3}''$ pieces, and one $11^{3}\!/_{4}''$ piece can be cut from it? Disregard saw kerfs.

_____ **31.** A jogger ran $2^{1}\!/_{2}$ miles Monday, $2^{1}\!/_{2}$ miles Tuesday, $2^{1}\!/_{4}$ miles Wednesday, $2^{3}\!/_{8}$ miles Thursday, and $1^{7}\!/_{8}$ miles Saturday. The jogger did not run Friday and Sunday. How many miles did the jogger run during the week?

_____ **32.** A landscape contractor has a stockpile containing 50 cu yd of topsoil. One job requires $11\frac{1}{2}$ cu yd and two small jobs require $4\frac{1}{4}$ cu yd each. How much topsoil is left in the stockpile?

_____ **33.** A secretary has $1\frac{1}{2}$ reams of $8\frac{1}{2}'' \times 11''$ paper in stock. Job A requires $\frac{3}{8}$ ream and Job B requires $\frac{1}{4}$ ream. How much of the original stock is remaining?

_____ **34.** Three pieces of steel rod measure $7\frac{1}{4}''$ each. Two other pieces of steel rod measure $5\frac{3}{8}''$ each. What is the total length of the steel rod?

_____ **35.** Ed uses $4\frac{3}{8}$ gal. of paint to paint an apartment. Doyle uses $4\frac{1}{2}$ gal. of paint to paint an identical apartment. How many gallons of paint did Ed and Doyle use?

Name _____ Date _____

_____ **1.** Change $^{10}\!/_{50}$ to higher terms using 2 as a multiplier.

_____ **2.** Find the LCD for $^{3}\!/_{4}$, $^{3}\!/_{8}$, $^{1}\!/_{2}$, $^{5}\!/_{16}$, and $^{1}\!/_{4}$.

_____ **3.** Find the LCD for $^{5}\!/_{8}$, $^{7}\!/_{32}$, $^{13}\!/_{64}$, $^{1}\!/_{16}$, and $^{3}\!/_{4}$.

Reduce the fractions to lowest possible terms.

_____ **4.** $^{10}\!/_{75}$

_____ **5.** $^{100}\!/_{120}$

Change the improper fractions to mixed numbers.

_____ **6.** $^{102}\!/_{8}$

_____ **7.** $^{46}\!/_{3}$

Change the mixed numbers to improper fractions.

_____ **8.** $35^{3}\!/_{5}$

_____ **9.** $7^{1}\!/_{10}$

_____ **10.** An electrical contractor bought a 1000′ reel of wire. One job required $125^{1}\!/_{2}′$ and $118^{3}\!/_{4}′$ of wire. Another job required $174^{2}\!/_{3}′$ and $236^{5}\!/_{6}′$. Find the length of wire remaining.

_____ **11.** $^{3}\!/_{8} + ^{3}\!/_{4} + ^{3}\!/_{16}$

_____ **12.** $^{5}\!/_{32} + ^{11}\!/_{32} + ^{19}\!/_{32}$

_____ **13.** $^{23}\!/_{32} - ^{5}\!/_{16}$

_____ **14.** $^{7}\!/_{8} - ^{3}\!/_{4}$

_____ **15.** $3\frac{5}{32} + 11\frac{1}{2}$

_____ **16.** $12\frac{1}{4} + 3\frac{1}{2}$

_____ **17.** $15\frac{15}{32} - 7\frac{7}{16}$

_____ **18.** $4\frac{3}{4} - 2\frac{3}{8}$

_____ **19.** A hardware store has $71\frac{1}{2}$ boxes of $\frac{1}{4}$-20 UNC bolts and $18\frac{1}{4}$ boxes of $\frac{1}{4}$-28 UNF bolts. How many boxes of $\frac{1}{4}''$ bolts does the store have?

_____ **20.** A carpenter cuts $5\frac{3}{8}''$ off a $16\frac{1}{2}''$ board. Disregarding the saw kerf, what is the length of the remaining board?

_____ **21.** A mason uses $1\frac{1}{4}$ cu yd of sand from a pile containing $4\frac{1}{2}$ cu yd. How much sand is left in the pile?

_____ **22.** Three tractor operators work $4\frac{1}{4}$ hr, $4\frac{1}{2}$ hr, and $4\frac{1}{2}$ hr to grade the right-of-way on a road project. What is the total number of hours required to grade the right-of-way?

_____ **23.** John uses $3\frac{5}{8}$ gal. of paint to paint a room. Fred uses $3\frac{1}{4}$ gal. of paint to paint an identical room. How much more paint did John use?

_____ **24.** A cabinetmaker uses $\frac{3}{8}$ box of door pulls for one job and $\frac{1}{4}$ box of door pulls for another job. How much of the box of door pulls remains?

FRACTIONS:
Multiplying and Dividing

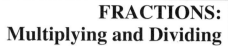

F ractions can be multiplied horizontally or vertically. Fractions are divided horizontally. Many fraction combinations, such as two fractions, two mixed numbers, a fraction and a whole number, etc., can be multiplied and divided. Each fraction combination follows a different rule.

MULTIPLYING FRACTIONS

Fractions can be multiplied horizontally or vertically. Horizontal placement of fractions is the most common because identification of the numerators and denominators is more obvious.

Fraction combinations that may be multiplied include two fractions, a fraction by a whole number, a mixed number by a whole number, and two mixed numbers. Each fraction combination follows a different rule.

Multiplying Two Fractions

To multiply two fractions, multiply the numerator of one fraction by the numerator of the other fraction. Follow the same procedure with the denominators. Reduce the answer if

required. See Figure 4-1. For example, to multiply $\frac{1}{4}$ by $\frac{1}{2}$ ($\frac{1}{4} \times \frac{1}{2}$), multiply the numerators and multiply the denominators ($\frac{1 \times 1}{4 \times 2} = \frac{1}{8}$).

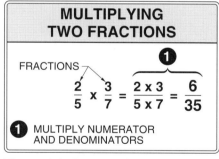

MULTIPLYING TWO FRACTIONS

FRACTIONS

$$\frac{2}{5} \times \frac{3}{7} = \frac{2 \times 3}{5 \times 7} = \frac{6}{35}$$

1 MULTIPLY NUMERATOR AND DENOMINATORS

Figure 4-1. To multiply two fractions, multiply numerators and denominators.

In many cases, cancellation can be used to simplify the multiplication process. For example, when multiplying $\frac{2}{5} \times \frac{1}{8}$, the 2 divides into both 2 (of the $\frac{2}{5}$) and 8 (of the $\frac{1}{8}$) an exact number of times

$(\frac{2 \div 2}{5} \times \frac{1}{8 \div 2} = \frac{1}{5} \times \frac{1}{4})$. No more cancellation is possible $(\frac{1}{5} \times \frac{1}{4} = \frac{1 \times 1}{5 \times 4} = \mathbf{\frac{1}{20}})$.

EXAMPLES

Multiplying Two Fractions

1. Multiply $\frac{3}{8}$ by $\frac{1}{8}$.

FRACTIONS

$$\frac{3}{8} \times \frac{1}{8} = \overbrace{\frac{3 \times 1}{8 \times 8}}^{❶} = \frac{3}{64}$$

❶ MULTIPLY NUMERATORS AND DENOMINATORS

① Multiply the numerators and denominators $(\frac{3 \times 1}{8 \times 8} = \mathbf{\frac{3}{64}})$.

2. Multiply $\frac{1}{2}$ by $\frac{1}{8}$.

$$\frac{1}{2} \times \frac{1}{8} = \overbrace{\frac{1 \times 1}{2 \times 8}}^{❶} = \frac{1}{16}$$

❶ MULTIPLY NUMERATORS AND DENOMINATORS

① Multiply the numerators and denominators $(\frac{1 \times 1}{2 \times 8} = \mathbf{\frac{1}{16}})$.

3. Multiply $\frac{2}{3}$ by $\frac{2}{6}$. Use cancellation.

$$\frac{2}{3} \times \frac{2}{6} = \overbrace{\frac{\cancel{2}^{1}}{3} \times \frac{2}{\cancel{6}_{3}}}^{❶} = \overbrace{\frac{1 \times 2}{3 \times 3}}^{❷} = \frac{2}{9}$$

❶ CANCEL

❷ MULTIPLY NUMERATORS AND DENOMINATORS

① Cancel the 2 (of the $\frac{2}{3}$) and the 6 (of the $\frac{2}{6}$) by dividing by 2 $(\frac{2 \div 2}{3} \times \frac{2}{6 \div 2} = \frac{1}{3} \times \frac{2}{3})$. No more cancellation is possible. ② Multiply the numerators and denominators $(\frac{1}{3} \times \frac{2}{3} = \frac{1 \times 2}{3 \times 3} = \mathbf{\frac{2}{9}})$.

PRACTICE PROBLEMS

Multiplying Two Fractions

1. $\frac{1}{2} \times \frac{1}{5}$
2. $\frac{1}{6} \times \frac{3}{7}$
3. $\frac{2}{3} \times \frac{4}{5}$
4. $\frac{5}{10} \times \frac{5}{6}$
5. $\frac{3}{32} \times \frac{8}{10}$
6. $\frac{7}{16} \times \frac{4}{5}$
7. $\frac{15}{16} \times \frac{4}{5}$
8. $\frac{8}{10} \times \frac{4}{7}$
9. An electrician has $\frac{3}{4}$ of a house to wire. In one day $\frac{2}{5}$ of the job is completed. Find how much of the house is completed that day.
10. An oxygen tank for oxyfuel welding is $\frac{4}{5}$ full. During the fabrication of a component, $\frac{2}{9}$

of the oxygen is used. How much of the $\frac{4}{5}$ full tank is used?

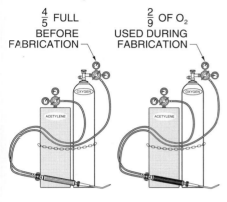

$\frac{4}{5}$ FULL
BEFORE
FABRICATION

$\frac{2}{9}$ OF O₂
USED DURING
FABRICATION

MULTIPLYING IMPROPER FRACTIONS

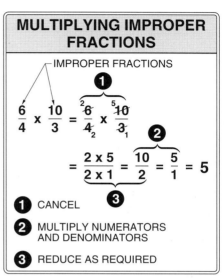

IMPROPER FRACTIONS

1

$$\frac{6}{4} \times \frac{10}{3} = \frac{\overset{2}{\cancel{6}}}{\underset{2}{\cancel{4}}} \times \frac{\overset{5}{\cancel{10}}}{\underset{1}{\cancel{3}}}$$

2

$$= \frac{2 \times 5}{2 \times 1} = \frac{10}{2} = \frac{5}{1} = 5$$

3

1 CANCEL

2 MULTIPLY NUMERATORS AND DENOMINATORS

3 REDUCE AS REQUIRED

Figure 4-2. To multiply improper fractions, use cancellation and multiply numerators and denominators.

Multiplying Improper Fractions

To multiply one improper fraction by another improper fraction, use cancellation if possible. Multiply the numerator of one fraction by the numerator of the other fraction. Follow the same procedure with the denominators. Reduce the answer if required. Change all answers from improper fractions to mixed numbers. See Figure 4-2.

For example, to multiply $\frac{18}{5} \times \frac{13}{12}$, cancel the 18 (of $\frac{18}{5}$) and the 12 (of $\frac{13}{12}$) by dividing by 6 ($\frac{18 \div 6}{5} \times \frac{13}{12 \div 6} = \frac{3}{5} \times \frac{13}{2}$). Multiply the numerators and denominators ($\frac{3 \times 13}{5 \times 2} = \frac{39}{10}$). Reduce as required ($\frac{39}{10} = \mathbf{3\frac{9}{10}}$).

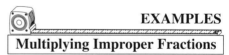

EXAMPLES

Multiplying Improper Fractions

1. Multiply $\frac{18}{3}$ by $\frac{24}{9}$.

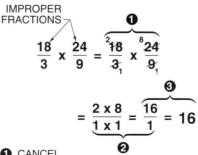

IMPROPER FRACTIONS

1

$$\frac{18}{3} \times \frac{24}{9} = \frac{\overset{2}{\cancel{18}}}{\underset{1}{\cancel{3}}} \times \frac{\overset{8}{\cancel{24}}}{\underset{1}{\cancel{9}}}$$

3

$$= \frac{2 \times 8}{1 \times 1} = \frac{16}{1} = 16$$

2

1 CANCEL

2 MULTIPLY NUMERATORS AND DENOMINATORS

3 REDUCE AS REQUIRED

① Cancel the 18 (of $^{18}\!/_3$) and the 9 (of $^{24}\!/_9$) by dividing by 9. Cancel the 3 (of $^{18}\!/_3$) and the 24 (of $^{24}\!/_9$) by dividing by 3 ($\dfrac{18 \div 9}{3 \div 3} \times \dfrac{24 \div 3}{9 \div 9} = {}^2\!/_1 \times {}^8\!/_1$). No more cancellation is possible. The problem now is $^2\!/_1 \times {}^8\!/_1$. ② Multiply the numerators and denominators ($\dfrac{2 \times 8}{1 \times 1} = {}^{16}\!/_1$). ③ Reduce as required ($^{16}\!/_1 = \mathbf{16}$).

2. Multiply $^{11}\!/_7$ by $^{15}\!/_4$.

$$\frac{11}{7} \times \frac{15}{4} = \overbrace{\frac{11 \times 15}{7 \times 4}}^{\text{❶}} = \underbrace{\frac{165}{28}}_{\text{❷}} = 5\frac{25}{28}$$

❶ MULTIPLY NUMERATORS AND DENOMINATORS
❷ REDUCE AS REQUIRED

No cancellation is possible. ① Multiply the numerators and denominators ($\dfrac{11 \times 15}{7 \times 4} = {}^{165}\!/_{28}$). ② Reduce as required ($^{165}\!/_{28} = \mathbf{5^{25}\!/_{28}}$).

PRACTICE PROBLEMS

Multiplying Improper Fractions

1. Multiply $^{21}\!/_3$ by $^{27}\!/_7$.
2. Multiply $^{23}\!/_{16}$ by $^{24}\!/_9$.

3. Multiply $^{11}\!/_{10}$ by $^5\!/_4$.
4. Multiply $^4\!/_3$ by $^{17}\!/_{12}$.
5. Multiply $^{84}\!/_{21}$ by $^7\!/_4$.

Multiplying More Than Two Fractions

To multiply more than two fractions, use cancellation if possible. Multiply the numerator of the first fraction by the numerator of the second fraction. The product is multiplied by the numerator of the third fraction and so on. Follow the same procedure with the denominators. Reduce the answer if required. Change all answers from improper fractions to mixed numbers. See Figure 4-3.

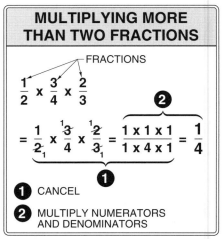

MULTIPLYING MORE THAN TWO FRACTIONS

❶ CANCEL
❷ MULTIPLY NUMERATORS AND DENOMINATORS

Figure 4-3. To multiply more than two fractions, use cancellation and multiply numerators and denominators.

For example, to multiply $^3/_5$ × $^6/_7$ × $^3/_{14}$, cancel the 6 (of $^6/_7$) and the 14 (of $^3/_{14}$) by dividing by 2 ($^3/_5$ × $\dfrac{6 \div 2}{7}$ × $\dfrac{3}{14 \div 2}$ = $^3/_5$ × $^3/_7$ × $^3/_7$). No more cancellation is possible. The problem now is $^3/_5$ × $^3/_7$ × $^3/_7$. Multiply the numerators and denominators ($\dfrac{3 \times 3 \times 3}{5 \times 7 \times 7}$ = $^{27}/_{245}$).

the 3 (of $^1/_3$) and the 6 (of $^6/_7$) by dividing by 3 ($\dfrac{4}{10 \div 2}$ × $\dfrac{1}{3 \div 3}$ × $\dfrac{4 \div 2}{5}$ × $\dfrac{6 \div 3}{7}$). No more cancellation is possible. The problem now is $^4/_5$ × $^1/_1$ × $^2/_5$ × $^2/_7$. ② Multiply the numerators and denominators ($\dfrac{4 \times 1 \times 2 \times 2}{5 \times 1 \times 5 \times 7}$ = $^{16}/_{175}$).

2. $^{34}/_3$ × $^9/_{17}$ × $^{15}/_2$ × $^6/_{30}$

Note: Proper and improper fractions can be multiplied. Use the same multiplication and cancellation process.

 EXAMPLES

Multiplying More Than Two Fractions

1. $^4/_{10}$ × $^1/_3$ × $^4/_5$ × $^6/_7$

⎧ ————FRACTIONS

❶⎨

$\dfrac{4}{10}$ x $\dfrac{1}{3}$ x $\dfrac{4}{5}$ x $\dfrac{6}{7}$ =

$\dfrac{4}{\cancel{10}_5}$ x $\dfrac{1}{3}$ x $\dfrac{\overset{2}{\cancel{4}}}{5}$ x $\dfrac{6}{7}$ =

$\dfrac{4}{5}$ x $\dfrac{1}{\cancel{3}_1}$ x $\dfrac{2}{5}$ x $\dfrac{\overset{2}{\cancel{6}}}{7}$ =

$\dfrac{4}{5}$ x $\dfrac{1}{1}$ x $\dfrac{2}{5}$ x $\dfrac{2}{7}$ =

$\dfrac{4 \times 1 \times 2 \times 2}{5 \times 1 \times 5 \times 7}$ = $\dfrac{16}{175}$

❷

❶ CANCEL
❷ MULTIPLY NUMERATORS AND DENOMINATORS

① Cancel the 10 (of $^4/_{10}$) and the 4 (of $^4/_5$) by dividing by 2. Cancel

❶⎨

$\dfrac{34}{3}$ x $\dfrac{9}{17}$ x $\dfrac{15}{2}$ x $\dfrac{6}{30}$ =

$\dfrac{\overset{2}{\cancel{34}}}{\cancel{3}_1}$ x $\dfrac{\overset{3}{\cancel{9}}}{\cancel{17}_1}$ x $\dfrac{15}{2}$ x $\dfrac{6}{30}$ =

$\dfrac{2}{1}$ x $\dfrac{3}{1}$ x $\dfrac{\overset{1}{\cancel{15}}}{\cancel{2}_1}$ x $\dfrac{\overset{3}{\cancel{6}}}{\cancel{30}_2}$ =

$\dfrac{\overset{1}{\cancel{2}}}{1}$ x $\dfrac{3}{1}$ x $\dfrac{1}{1}$ x $\dfrac{3}{\cancel{2}_1}$ =

$\dfrac{1}{1}$ x $\dfrac{3}{1}$ x $\dfrac{1}{1}$ x $\dfrac{3}{1}$ =

$\dfrac{1 \times 3 \times 1 \times 3}{1 \times 1 \times 1 \times 1}$ = $\dfrac{9}{1}$ = 9

❷ ❸

❶ CANCEL
❷ MULTIPLY NUMERATORS AND DENOMINATORS
❸ REDUCE AS REQUIRED

① Cancel the 34 (of $^{34}\!/_3$) and the 17 (of $^9\!/_{17}$) by dividing by 17. Cancel the 3 (of $^{34}\!/_3$) and the 9 (of $^9\!/_{17}$) by dividing by 3. Cancel the 15 (of $^{15}\!/_2$) and the 30 (of $^6\!/_{30}$) by dividing by 15. Cancel the 2 (of $^{15}\!/_2$) and the 6 (of $^6\!/_{30}$) by dividing by 2 ($\dfrac{34 \div 17}{3 \div 3} \times \dfrac{9 \div 3}{17 \div 17} \times \dfrac{15 \div 15}{2 \div 2}$ $\times \dfrac{6 \div 2}{30 \div 15}$). The problem now is $^2\!/_1 \times {}^3\!/_1 \times {}^1\!/_1 \times {}^3\!/_2$. Cancel the 2s produced by cancellation by dividing by 2. No more cancellation is possible. The problem now is $^1\!/_1 \times {}^3\!/_1 \times {}^1\!/_1 \times {}^3\!/_1$. ② Multiply the numerators and denominators ($\dfrac{1 \times 3 \times 1 \times 3}{1 \times 1 \times 1 \times 1} = {}^9\!/_1$). ③ Reduce as required ($^9\!/_1 = \mathbf{9}$).

PRACTICE PROBLEMS

Multiplying More Than Two Fractions

1. $^3\!/_5 \times {}^1\!/_4 \times {}^3\!/_4 \times {}^3\!/_{16}$
2. $^{35}\!/_{10} \times {}^2\!/_7 \times {}^{33}\!/_4 \times {}^{16}\!/_{11}$
3. $^3\!/_{10} \times {}^5\!/_9 \times {}^9\!/_4 \times {}^{12}\!/_{10}$
4. $^1\!/_{10} \times {}^2\!/_1 \times {}^3\!/_4 \times {}^{16}\!/_9$
5. $^5\!/_{10} \times {}^{24}\!/_3 \times {}^9\!/_5 \times {}^{11}\!/_4$

Multiplying a Fraction by Whole Number

To multiply a fraction and any whole number, multiply the numerator of the fraction by the whole number and place that product over the denominator. Reduce the answer if required. If necessary, change the answer from an improper fraction to a mixed number. See Figure 4-4. For example, to multiply $^1\!/_8$ by 3, multiply 1 (numerator) by 3 (whole number) and place 3 (product) over the denominator ($\dfrac{1 \times 3}{8} = {}^3\!/_8$).

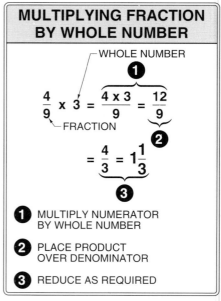

Figure 4-4. To multiply a fraction and a whole number, multiply the numerator by the whole number and place the product over the denominator.

EXAMPLES

Multiplying a Fraction by Whole Number

1. Multiply $\frac{3}{25}$ by 6.

WHOLE NUMBER

$$\frac{3}{25} \times 6 = \frac{3 \times 6}{25} = \frac{18}{25}$$

❶
❷
FRACTION

❶ MULTIPLY NUMERATOR BY WHOLE NUMBER

❷ PLACE PRODUCT OVER DENOMINATOR

① Multiply 3 (numerator) by 6 (whole number) ($3 \times 6 = 18$). ② Place 18 (product) over the denominator ($\frac{3 \times 6}{25} = \frac{18}{25}$).

2. Multiply $\frac{5}{10}$ by 10.

❶

$$\frac{5}{10} \times 10 = \frac{5 \times 10}{10} = \frac{50}{10} = 5$$

❷
❸

❶ MULTIPLY NUMERATOR BY WHOLE NUMBER

❷ PLACE PRODUCT OVER DENOMINATOR

❸ REDUCE AS REQUIRED

① Multiply 5 (numerator) by 10 (whole number) ($5 \times 10 = 50$). ② Place 50 (product) over the denominator ($\frac{50}{10}$). ③ Reduce as required ($50 \div 10 = 5$).

PRACTICE PROBLEMS

Multiplying a Fraction by Whole Number

1. Multiply $\frac{2}{9}$ by 2.
2. Multiply $\frac{1}{10}$ by 3.
3. Multiply $\frac{1}{20}$ by 7.
4. Multiply $\frac{3}{11}$ by 3.
5. Multiply $\frac{4}{5}$ by 12.
6. Multiply $\frac{3}{9}$ by 33.
7. Multiply $\frac{4}{7}$ by 11.
8. A motor delivers $\frac{8}{9}$ of the power it receives. How much power does the motor deliver if it receives 25 HP?
9. Find the total amount of paint in the cans.

1 GAL. CANS $\frac{3}{5}$ FULL

10. A machinist takes $\frac{1}{6}$ of an hour to machine a component. How long does the machinist take to machine 25 components?

Cancellation Method

The cancellation method is a quick and easy method of multiplying a fraction by a whole number. See Figure 4-5. For example, $\frac{5}{10} \times 10$

can be simplified by cancelling the two 10s by dividing by 10. After cancelling, the problem is $\frac{5}{1} \times 1$, or $\frac{5}{1} = 5$. Reduce the answer if required. If necessary, change the answer from an improper fraction to a mixed number.

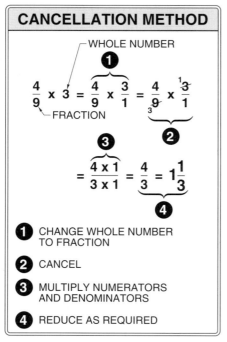

Figure 4-5. Cancellation simplifies multiplying fractions and whole numbers.

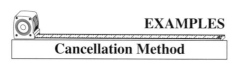

EXAMPLES

Cancellation Method

1. Multiply $^{11}\!/_{72} \times 18$.

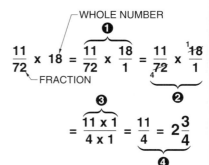

❶ CHANGE WHOLE NUMBER TO FRACTION

❷ CANCEL

❸ MULTIPLY NUMERATORS AND DENOMINATORS

❹ REDUCE AS REQUIRED

① Change the whole number (18) to a fraction ($^{18}\!/_{1}$). ② Cancel the 72 (of $^{11}\!/_{72}$) and the 18 (whole number) by dividing by 18 ($\frac{11 \div 72}{18} \times \frac{18 \div 18}{1}$). No more cancellation is possible. The problem now is $^{11}\!/_{4} \times 1$. ③ Multiply 11 (numerator) by 1 (whole number) and place the 11 (product) over the denominator ($\frac{11 \times 1}{4} = {}^{11}\!/_{4}$). The fraction $^{11}\!/_{4}$ is an improper fraction. ④ Reduce the improper fraction by dividing the numerator by the denominator ($11 \div 4 = 2\frac{3}{4}$).

PRACTICE PROBLEMS

Cancellation Method

1. Multiply $^{4}\!/_{5}$ by 35.

2. Multiply $^{5}\!/_{84}$ by 84.

3. Multiply $\frac{7}{50}$ by 10.

4. Multiply $\frac{49}{65}$ by 13.

5. Multiply $\frac{5}{12}$ by 132.

6. Multiply $\frac{7}{64}$ by 96.

7. A masonry contractor has 420 common bricks to lay. After 5 hours, $\frac{3}{4}$ of the bricks are laid. How many bricks are left to lay?

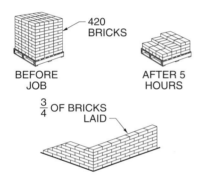

8. A carpenter figures approximately 300 sq ft of forms can be erected in 8 hours. If $\frac{1}{3}$ of the forms have been erected, how many square feet of forms remain to be erected?

9. A carpenter has 6 lb of nails and uses $\frac{5}{8}$ of them. How many pounds of nails are used?

10. Four hundred components are turned on an automatic lathe. An inspector rejects $\frac{1}{20}$ of them due to faulty material. How many components are rejected?

Multiplying a Mixed Number by Whole Number

To multiply a mixed number and a whole number, use cancellation if possible. Multiply the fraction of the mixed number by the whole number. Multiply the whole numbers and add the two products. Reduce the answer if required. If necessary, change the answer from an improper fraction to a mixed number. See Figure 4-6.

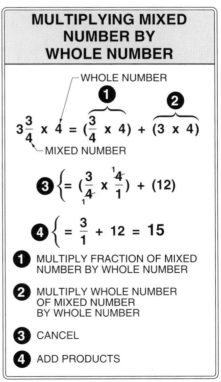

Figure 4-6. To multiply a mixed number and a whole number, multiply the fraction of the mixed number by the whole number, multiply whole numbers, and add products.

For example, to multiply $3\frac{3}{7}$ × 5, multiply $\frac{3}{7}$ (fraction of the mixed number) by 5 and reduce ($\frac{3}{7}$ × 5 = $\frac{15}{7}$ = 15 ÷ 7 = $2\frac{1}{7}$). Multiply the whole numbers (3 × 5 = 15). Add the two products ($2\frac{1}{7}$ + 15 = **$17\frac{1}{7}$**).

 EXAMPLES

Multiplying a Mixed Number by Whole Number

1. Multiply $4\frac{7}{8}$ by 3.

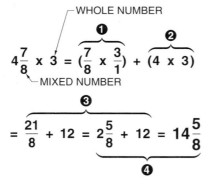

❶ MULTIPLY FRACTION OF MIXED NUMBER BY WHOLE NUMBER
❷ MULTIPLY WHOLE NUMBER OF MIXED NUMBER BY WHOLE NUMBER
❸ REDUCE AS REQUIRED
❹ ADD PRODUCTS

① Multiply $\frac{7}{8}$ (fraction of the mixed number) by 3 ($\frac{7}{8}$ × 3 = $\frac{21}{8}$). ② Multiply the whole numbers (4 × 3 = 12). ③ Reduce (21 ÷ 8 = $2\frac{5}{8}$). ④ Add the two products ($2\frac{5}{8}$ + 12 = **$14\frac{5}{8}$**).

PRACTICE PROBLEMS

Multiplying a Mixed Number by Whole Number

1. Multiply $81\frac{6}{7}$ by 49.
2. Multiply $35\frac{7}{8}$ by 36.
3. Multiply $47\frac{7}{10}$ by 65.
4. Multiply $13\frac{3}{5}$ by 22.
5. Multiply $3\frac{14}{19}$ by 35.
6. Multiply $35\frac{6}{71}$ by 5.
7. A trim carpenter needs 4 pieces of crown molding $13\frac{1}{2}'$ long to complete a job. How many lineal feet of crown molding are needed?

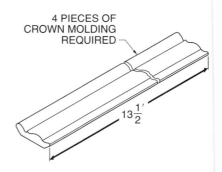

4 PIECES OF CROWN MOLDING REQUIRED

$13\frac{1}{2}'$

8. A welder takes $7\frac{3}{10}$ minutes to make a pipe weld. How much time is required to make 55 pipe welds?
9. What is the total rise of the stairs in the Stair Plan?
10. What is the total run of the stairs in the Stair Plan?

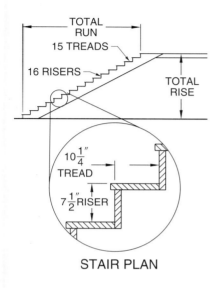

STAIR PLAN

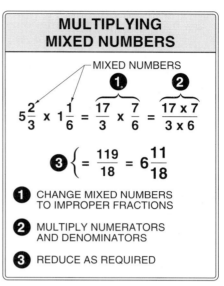

Figure 4-7. To multiply mixed numbers, change the mixed numbers to improper fractions and multiply.

Multiplying Mixed Numbers

To multiply one mixed number by another mixed number, change both mixed numbers to improper fractions, use cancellation if possible, and then multiply. See Figure 4-7. For example, to multiply $3\frac{1}{4}$ by $4\frac{1}{2}$, change $3\frac{1}{4}$ to an improper fraction by multiplying 3 by the denominator 4 ($3 \times 4 = 12$). Add the numerator 1 and place the sum over the denominator ($12 + \frac{1}{4} = \frac{13}{4}$). Change $4\frac{1}{2}$ to an improper fraction by multiplying 4 by the denominator 2 ($4 \times 2 = 8$). Add the numerator 1 and place the sum over the denominator ($8 + \frac{1}{2} = \frac{9}{2}$). Multiply the improper fractions ($\frac{13}{4} \times \frac{9}{2} = \frac{117}{8} = 117 \div 8 = \textbf{14}\frac{5}{8}$).

EXAMPLES

Multiplying Mixed Numbers

1. $9\frac{1}{3} \times 7\frac{4}{5}$

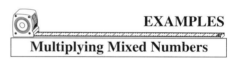

① Change the $9\frac{1}{3}$ to an improper fraction by multiplying 9 by the denominator 3 ($9 \times 3 = 27$). Add the numerator 1 to 27 and place the sum over 3 ($27 + \frac{1}{3} = \frac{28}{3}$). Change the $7\frac{4}{5}$ to an improper fraction by multiplying 7 by the denominator 5 ($7 \times 5 = 35$). Add the numerator 4 to 35 and place the sum over 5 ($35 + \frac{4}{5} = \frac{39}{5}$).

② Multiply the improper fractions ($\frac{28}{3} \times \frac{39}{5} = \dfrac{28 \times 39}{3 \times 5} = \frac{1092}{15}$).

③ Reduce as required ($1092 \div 15 = 72\frac{12}{15} = \mathbf{72\frac{4}{5}}$).

PRACTICE PROBLEMS

| **Multiplying Mixed Numbers** |

1. Multiply $28\frac{4}{9}$ by $16\frac{3}{4}$.
2. Multiply $38\frac{1}{2}$ by $12\frac{1}{2}$.
3. Multiply $51\frac{4}{5}$ by $72\frac{7}{9}$.
4. Multiply $8\frac{1}{2}$ by $3\frac{2}{3}$.
5. Multiply $12\frac{2}{7}$ by $28\frac{1}{3}$.
6. Multiply $3\frac{1}{5}$ by $2\frac{2}{3}$.
7. Multiply $33\frac{1}{5}$ by $20\frac{1}{2}$.
8. Multiply $17\frac{3}{10}$ by $5\frac{3}{5}$.
9. One gallon of water weighs approximately $8\frac{1}{3}$ lb. How much does $25\frac{3}{4}$ gal. of water weigh?

10. An experienced trim carpenter applied panelling and molding to panelled walls. An area of $700\frac{7}{12}$ sq ft is completed in one $7\frac{1}{3}$ hour shift. How many square feet will the trim carpenter complete in $2\frac{1}{2}$ shifts?

DIVIDING FRACTIONS

Fractions are divided horizontally. Fraction combinations that may be divided include two fractions, a fraction by a whole number, a mixed number by a whole number, a whole number by a fraction, and two mixed numbers. Each fraction combination follows a different rule.

Dividing Two Fractions

To divide one fraction by another fraction, invert the divisor fraction. Use cancellation if possible. Then multiply the numerator by the numerator and the denominator by the denominator. Reduce the answer as required. See Figure 4-8.

For example, to divide $\frac{3}{8}$ by $\frac{1}{4}$, invert $\frac{1}{4}$ (divisor fraction) and multiply by $\frac{3}{8}$ ($\frac{3}{8} \times \frac{4}{1} = \dfrac{3 \times 4}{8 \times 1} = \frac{12}{8}$). Reduce as required ($\frac{12}{8} = 12 \div 8 = 1\frac{4}{8} = \mathbf{1\frac{1}{2}}$).

DIVIDING TWO FRACTIONS

DIVISOR FACTOR

❶   ❷

$$\frac{4}{9} \div \frac{3}{32} = \frac{4}{9} \times \frac{32}{3} = \frac{4 \times 32}{9 \times 3}$$

❸

$$= \frac{128}{27} = 4\frac{20}{27}$$

❶ INVERT DIVISOR FRACTION

❷ MULTIPLY NUMERATORS AND DENOMINATORS

❸ REDUCE AS REQUIRED

Figure 4-8. To divide one fraction by another fraction, invert the divisor fraction and multiply numerators and denominators.

EXAMPLES

Dividing Two Fractions

1. Divide ⅔ by ¾.

DIVISOR FRACTION

❶   ❷

$$\frac{2}{3} \div \frac{3}{4} = \frac{2}{3} \times \frac{4}{3} = \frac{2 \times 4}{3 \times 3} = \frac{8}{9}$$

❶ INVERT DIVISOR FRACTION

❷ MULTIPLY NUMERATORS AND DENOMINATORS

① Invert ¾. ② Multiply ⅔ by ⁴⁄₃ $(⅔ \times ⁴⁄₃ = \frac{2 \times 4}{3 \times 3} = ⁸⁄₉)$.

2. Divide ⅞ by ⅔.

❶   ❹

$$\frac{7}{9} \div \frac{2}{3} = \frac{7}{\overset{3}{9}} \times \frac{\overset{1}{3}}{2} = \frac{7 \times 1}{3 \times 2} = \frac{7}{6} = 1\frac{1}{6}$$

❷   ❸

❶ INVERT DIVISOR FRACTION

❷ CANCEL

❸ MULTIPLY NUMERATORS AND DENOMINATORS

❹ REDUCE AS REQUIRED

① Invert ⅔. ② Cancel ⁷⁄₉ × ³⁄₂ = $\frac{7 \times 1}{3 \times 2}$. ③ Multiply $\frac{7 \times 1}{3 \times 2} = ⁷⁄₆$. ④ Reduce as required (⁷⁄₆ = **1⅙**).

PRACTICE PROBLEMS

Dividing Two Fractions

1. ¹³⁄₁₅ ÷ ⅔

2. ²⁵⁄₃₅ ÷ ¼

3. ¹¹⁄₁₅ ÷ ½

4. ⅜ ÷ ⅔

5. ¹⁵⁄₁₆ ÷ ¼

6. ⅘ ÷ ⅝

7. ⅝ ÷ ¼

8. ⁹⁄₁₆ ÷ ¼

9. A painter has ⅞ gal. of stain. An interior door requires ⅕ gal. of stain to finish. How many interior doors can be finished?

10. A scrap yard has ⁹⁄₁₀ t of scrap metal to be distributed into ¼ t bins. How many full bins are needed to distribute the metal?

Dividing a Fraction by Whole Number

To divide a fraction by a whole number, multiply the denominator of the fraction by the whole number. Then use cancellation if possible, carry the numerator, and reduce as required. See Figure 4-9.

For example, to divide $\frac{3}{8}$ by 4, multiply 8 (denominator) by 4 (whole number) and carry the 3 ($\frac{3}{8 \times 4} = \frac{3}{32}$).

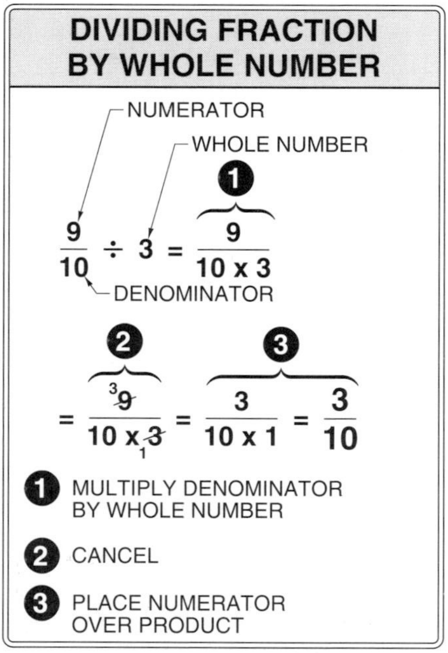

DIVIDING FRACTION BY WHOLE NUMBER

NUMERATOR

WHOLE NUMBER

❶

$$\frac{9}{10} \div 3 = \frac{9}{10 \times 3}$$

DENOMINATOR

❷   ❸

$$= \frac{\overset{3}{\cancel{9}}}{10 \times \underset{1}{\cancel{3}}} = \frac{3}{10 \times 1} = \frac{3}{10}$$

❶ MULTIPLY DENOMINATOR BY WHOLE NUMBER

❷ CANCEL

❸ PLACE NUMERATOR OVER PRODUCT

Figure 4-9. To divide a fraction by a whole number, multiply the denominator by the whole number and carry the numerator.

EXAMPLES

Dividing a Fraction by Whole Number

1. Divide $\frac{3}{5}$ by 2.

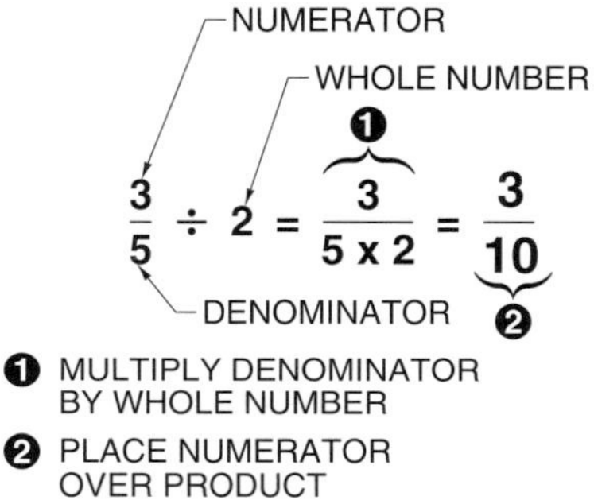

NUMERATOR

WHOLE NUMBER

❶

$$\frac{3}{5} \div 2 = \frac{3}{5 \times 2} = \frac{3}{10}$$

DENOMINATOR   ❷

❶ MULTIPLY DENOMINATOR BY WHOLE NUMBER

❷ PLACE NUMERATOR OVER PRODUCT

① Multiply 5 by 2. ② Carry the 3 ($\frac{3}{5 \times 2} = \frac{3}{10}$).

2. Divide $\frac{4}{5}$ by 4.

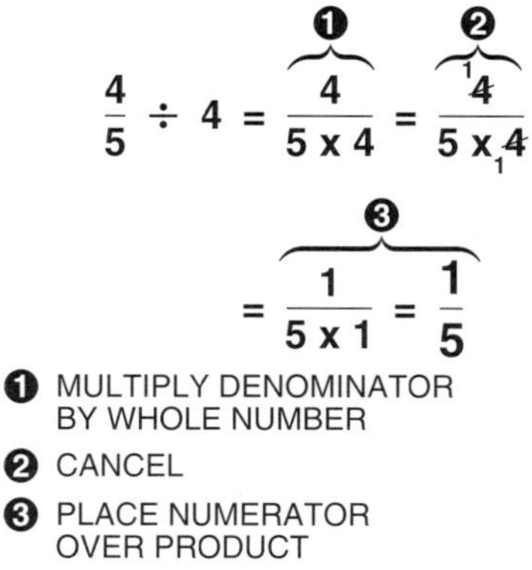

❶   ❷

$$\frac{4}{5} \div 4 = \frac{4}{5 \times 4} = \frac{\overset{1}{\cancel{4}}}{5 \times \underset{1}{\cancel{4}}}$$

❸

$$= \frac{1}{5 \times 1} = \frac{1}{5}$$

❶ MULTIPLY DENOMINATOR BY WHOLE NUMBER

❷ CANCEL

❸ PLACE NUMERATOR OVER PRODUCT

① Multiply 5 by 4. ② Cancel the 4 (numerator) and 4 (whole number) by dividing by 4. No more cancellation is possible. The problem now is $\frac{1}{5} \times 1$. ③ Carry the 1 ($\frac{1}{5} \times 1 = \frac{1}{5}$).

nominator) by 3 (whole number) and carry the 23 ($\frac{23}{8 \times 3} = \frac{23}{24}$).

DIVIDING MIXED NUMBER BY WHOLE NUMBER

MIXED NUMBER

WHOLE NUMBER

1 **2**

$8\frac{2}{3} \div 4 = \dfrac{\overbrace{26}}{3} \div 4 = \dfrac{\overbrace{26}}{3 \times 4}$

3 **5**

$= \dfrac{\overset{13}{\overbrace{26}}}{3 \times \underset{2}{4}} = \dfrac{13}{3 \times 2} = \underbrace{\dfrac{13}{6}} = 2\frac{1}{6}$

4

1 CHANGE MIXED NUMBER TO IMPROPER FRACTION

2 MULTIPLY DENOMINATOR BY WHOLE NUMBER

3 CANCEL

4 CARRY NUMERATOR

5 REDUCE AS REQUIRED

Figure 4-10. To divide a mixed number by a whole number, change to improper fraction, multiply the denominator by the whole number, and carry the numerator.

PRACTICE PROBLEMS

Dividing a Fraction by Whole Number

1. $^{30}/_{32} \div 5$
2. $^{33}/_{55} \div 11$
3. $^{31}/_{32} \div 3$
4. $^5/_8 \div 7$
5. $^7/_{20} \div 4$
6. $^1/_2 \div 7$
7. $^3/_8 \div 16$
8. A painter paints three doors with $^3/_4$ gal. of paint. How much paint is needed to paint one door?

Dividing a Mixed Number by Whole Number

To divide a mixed number by a whole number, change the mixed number to an improper fraction. Multiply the denominator of the fraction by the whole number, use cancellation if possible, carry the numerator, and reduce as required. See Figure 4-10.

For example, to divide $2\frac{7}{8}$ by 3, change the $2\frac{7}{8}$ to an improper fraction ($2\frac{7}{8} = \frac{23}{8}$). Multiply 8 (de-

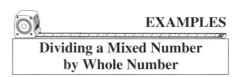

EXAMPLES

Dividing a Mixed Number by Whole Number

1. $3\frac{3}{8} \div 3$

MIXED NUMBER

WHOLE NUMBER

❶ ❷

$$3\frac{3}{8} \div 3 = \frac{27}{8} \div 3 = \frac{27}{8 \times 3}$$

❸ ❺

$$= \frac{^{9}27}{8 \times \underset{1}{3}} = \frac{9}{8 \times 1} = \frac{9}{8} = 1\frac{1}{8}$$

❹

❶ CHANGE MIXED NUMBER
   TO IMPROPER FRACTION
❷ MULTIPLY DENOMINATOR
   BY WHOLE NUMBER
❸ CANCEL
❹ CARRY NUMERATOR
❺ REDUCE AS REQUIRED

① Change the $3\frac{3}{8}$ to an improper fraction ($3\frac{3}{8} = {}^{27}\!/_{8}$). ② Multiply 8 by 3. ③ Cancel the 27 and 3 by dividing by 3. No more cancellation is possible. The problem now is $\frac{9}{8} \times 1$. ④ Carry the 9 ($\frac{9}{8} \times 1 = \frac{9}{8}$). ⑤ Reduce $\frac{9}{8}$ to a mixed number ($\frac{9}{8} = 9 \div 8 = \mathbf{1\frac{1}{8}}$).

2. $4\frac{2}{9} \div 5$

$$4\frac{2}{9} \div 5 = \frac{38}{9} \div 5 \Big\} \ \mathbf{❶}$$

$$= \frac{38}{9 \times 5} = \frac{38}{45} \Big\} \ \mathbf{❸}$$

❷

❶ CHANGE MIXED NUMBER
   TO IMPROPER FRACTION
❷ MULTIPLY DENOMINATOR
   BY WHOLE NUMBER
❸ CARRY NUMERATOR

① Change the $4\frac{2}{9}$ to an improper fraction ($4\frac{2}{9} = {}^{38}\!/_{9}$). ② Multiply 9 by 5. ③ Carry the 38 ($\frac{38}{9 \times 5} = {}^{38}\!/_{45}$).

PRACTICE PROBLEMS

Dividing a Mixed Number by Whole Number

1. $250\frac{1}{2} \div 5$
2. $333\frac{1}{3} \div 3$
3. $789\frac{2}{5} \div 4$
4. $877\frac{1}{7} \div 7$
5. $724\frac{2}{9} \div 8$
6. $5\frac{1}{2} \div 2$
7. $17\frac{7}{9} \div 8$
8. A concrete mixer mixes 586 $\frac{2}{3}$ cu ft of concrete in 5 hours. At the same rate, how much concrete is mixed in 1 hour?
9. What is the width of each step of the Step Pulley?

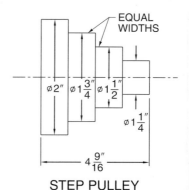

STEP PULLEY

10. A $43\frac{1}{2}''$ piece of pipe is cut into three pieces. What is the length of each piece? Disregard the saw kerf.

Dividing a Whole Number by Fraction

To divide a whole number by a fraction, change the whole number to fraction form and invert the divisor fraction. Then use cancellation if possible, multiply the numerators and the denominators, and reduce as required. See Figure 4-11.

For example, to divide 12 by $\frac{3}{4}$, change the 12 (whole number) to $\frac{12}{1}$ (fraction form). Invert $\frac{3}{4}$ (divisor fraction). Cancel 12 ($\frac{12}{1}$) and 3 ($\frac{4}{3}$) by dividing by 3 ($\frac{12 \div 3}{1} \times \frac{4}{3 \div 3} = \frac{4}{1} \times \frac{4}{1}$). No more cancellation is possible. Multiply the numerators and denominators ($\frac{4 \times 4}{1 \times 1} = \frac{16}{1}$). Reduce as required ($\frac{16}{1} = 16 \div 1 = $ **16**).

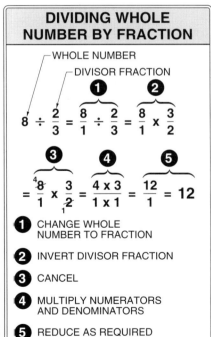

DIVIDING WHOLE NUMBER BY FRACTION

WHOLE NUMBER
DIVISOR FRACTION

① **②**

$$8 \div \frac{2}{3} = \frac{8}{1} \div \frac{2}{3} = \frac{8}{1} \times \frac{3}{2}$$

③ **④** **⑤**

$$= \frac{\overset{4}{\cancel{8}}}{1} \times \frac{3}{\underset{1}{\cancel{2}}} = \frac{4 \times 3}{1 \times 1} = \frac{12}{1} = 12$$

① CHANGE WHOLE NUMBER TO FRACTION

② INVERT DIVISOR FRACTION

③ CANCEL

④ MULTIPLY NUMERATORS AND DENOMINATORS

⑤ REDUCE AS REQUIRED

Figure 4-11. To divide a whole number by a fraction, change to fraction form, invert divisor fraction, and multiply numerators and denominators.

EXAMPLES

Dividing a Whole Number by Fraction

1. Divide 10 by $\frac{2}{5}$.

WHOLE NUMBER
DIVISOR FRACTION

① **②**

$$10 \div \frac{2}{5} = \frac{10}{1} \div \frac{2}{5} = \frac{10}{1} \times \frac{5}{2}$$

③ **④** **⑤**

$$= \frac{\overset{5}{\cancel{10}}}{1} \times \frac{5}{\underset{1}{\cancel{2}}} = \frac{5 \times 5}{1 \times 1} = \frac{25}{1} = 25$$

① CHANGE WHOLE NUMBER TO FRACTION

② INVERT DIVISOR FRACTION

③ CANCEL

④ MULTIPLY NUMERATORS AND DENOMINATORS

⑤ REDUCE AS REQUIRED

① Change the 10 to $^{10}/_1$. ② Invert $^2/_5$. ③ Cancel 10 ($^{10}/_1$) and 2 ($^5/_2$) by dividing by 2 ($\dfrac{10 \div 2}{1} \times \dfrac{5}{2 \div 2} =$ $^5/_1 \times ^5/_1$). No more cancellation is possible. ④ Multiply the numerators and denominators ($\dfrac{5 \times 5}{1 \times 1} = ^{25}/_1$). ⑤ Reduce as required ($^{25}/_1 = 25 \div 1 = 25$).

2. Divide 36 by $^9/_{13}$.

$$36 \div \dfrac{\overset{\textbf{①}}{9}}{13} = \dfrac{\overset{\textbf{①}}{36}}{1} \div \dfrac{9}{13} = \dfrac{36}{1} \times \dfrac{\overset{\textbf{②}}{13}}{9}$$

$$= \dfrac{\overset{\textbf{③}}{^4 36}}{1} \times \dfrac{13}{\underset{1}{9}} = \dfrac{\overset{\textbf{④}}{4 \times 13}}{1 \times 1} = \dfrac{\overset{\textbf{⑤}}{52}}{1} = 52$$

① CHANGE WHOLE NUMBER TO FRACTION
② INVERT DIVISOR FRACTION
③ CANCEL
④ MULTIPLY NUMERATORS AND DENOMINATORS
⑤ REDUCE AS REQUIRED

① Change the 36 to $^{36}/_1$. ② Invert $^9/_{13}$. ③ Cancel 36 ($^{36}/_1$) and 9 ($^{13}/_9$) by dividing by 9 ($\dfrac{36 \div 9}{1} \times \dfrac{13}{9 \div 9} =$ $^4/_1 \times ^{13}/_1$). No more cancellation is possible. ④ Multiply the numerators and denominators

($\dfrac{4 \times 13}{1 \times 1} = ^{52}/_1$). ⑤ Reduce as required ($^{52}/_1 = 52 \div 1 = 52$).

PRACTICE PROBLEMS

Dividing a Whole Number by Fraction

1. $25 \div ^5/_7$
2. $42 \div ^6/_7$
3. $26 \div ^4/_9$
4. $65 \div ^{13}/_{15}$
5. $81 \div ^9/_{11}$

Dividing Mixed Numbers

To divide a mixed number by a mixed number, change mixed numbers to improper fractions and invert the divisor fraction. Then use cancellation if possible, multiply the numerators and denominators, and reduce as required. See Figure 4-12.

For example, to divide $12^1/_2$ by $3^1/_8$, change $12^1/_2$ and $3^1/_8$ to improper fractions ($12^1/_2 = ^{25}/_2$ and $3^1/_8 = ^{25}/_8$). Invert $^{25}/_8$ (divisor fraction) and multiply with $^{25}/_2$. Cancel 25 ($^{25}/_2$) and 25 ($^8/_{25}$) by dividing by 25. Cancel 2 ($^{25}/_2$) and 8 ($^8/_{25}$) by dividing by 2 ($\dfrac{25 \div 25}{2 \div 2} \times \dfrac{8 \div 2}{25 \div 25} = ^1/_1 \times$ $^4/_1$). No more cancellation is possible.

Multiply the numerators and denomi-

nators $(\dfrac{1 \times 4}{1} \times 1 = \sqrt[4]{1} = 4 \div 1 = \mathbf{4})$.

DIVIDING MIXED NUMBERS

MIXED NUMBERS

❶   **❷**

$1\dfrac{1}{3} \div 1\dfrac{1}{9} = \dfrac{4}{3} \div \dfrac{10}{9} = \dfrac{4}{3} \times \dfrac{9}{10}$

❸   **❹**   **❺**

$= \dfrac{\overset{2}{4}}{\underset{1}{3}} \times \dfrac{\overset{3}{9}}{\underset{5}{10}} = \dfrac{2 \times 3}{1 \times 5} = \dfrac{6}{5} = 1\dfrac{1}{5}$

❶ CHANGE MIXED NUMBERS TO IMPROPER FRACTION

❷ INVERT DIVISOR FRACTION

❸ CANCEL

❹ MULTIPLY NUMERATORS AND DENOMINATORS

❺ REDUCE AS REQUIRED

Figure 4-12. To divide a mixed number by a mixed number, change to improper fractions, invert the divisor fraction, and multiply numerators and denominators.

 EXAMPLES

Dividing Mixed Numbers

1. $7\frac{2}{9} \div 4\frac{1}{3}$

❶   **❷**

$7\dfrac{2}{9} \div 4\dfrac{1}{3} = \dfrac{65}{9} \div \dfrac{13}{3} = \dfrac{65}{9} \times \dfrac{3}{13}$

MIXED NUMBERS

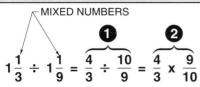

❸   **❹**   **❺**

$= \dfrac{\overset{5}{65}}{\underset{3}{9}} \times \dfrac{\overset{1}{3}}{\underset{1}{13}} = \dfrac{5 \times 1}{3 \times 1} = \dfrac{5}{3} = 1\dfrac{2}{3}$

❶ CHANGE MIXED NUMBERS TO IMPROPER FRACTIONS

❷ INVERT DIVISOR FRACTION

❸ CANCEL

❹ MULTIPLY NUMERATORS AND DENOMINATORS

❺ REDUCE AS REQUIRED

① Change $7\frac{2}{9}$ and $4\frac{1}{3}$ to improper fractions ($7\frac{2}{9} = \frac{65}{9}$ and $4\frac{1}{3} = \frac{13}{3}$). ② Invert $\frac{13}{3}$ and multiply by $\frac{65}{9}$. ③ Cancel $\frac{65}{9} \times \frac{3}{13}$ to $\dfrac{5 \times 1}{3 \times 1}$. ④ Multiply the numerators and denominators ($\dfrac{5 \times 1}{3 \times 1} = \frac{5}{3}$). ⑤ Reduce as required ($\frac{5}{3} = 5 \div 3 = \mathbf{1\frac{2}{3}}$).

2. $8\frac{3}{4} \div 7\frac{1}{2}$

❶   **❷**

$8\dfrac{3}{4} \div 7\dfrac{1}{2} = \dfrac{35}{4} \div \dfrac{15}{2} = \dfrac{35}{4} \times \dfrac{2}{15}$

❸   **❹**   **❺**

$= \dfrac{\overset{7}{35}}{\underset{2}{4}} \times \dfrac{\overset{1}{2}}{\underset{3}{15}} = \dfrac{7 \times 1}{2 \times 3} = \dfrac{7}{6} = 1\dfrac{1}{6}$

❶ CHANGE MIXED NUMBERS TO IMPROPER FRACTIONS

❷ INVERT DIVISOR FRACTION

❸ CANCEL

❹ MULTIPLY NUMERATORS AND DENOMINATORS

❺ REDUCE AS REQUIRED

① Change $8\frac{3}{4}$ and $7\frac{1}{2}$ to improper fractions ($8\frac{3}{4} = \frac{35}{4}$ and $7\frac{1}{2} = \frac{15}{2}$). ② Invert $\frac{15}{2}$. ③ Cancel 35 ($\frac{35}{4}$) and 15 ($\frac{2}{15}$) by dividing by 5. Cancel 4 ($\frac{35}{4}$) and 2 ($\frac{2}{15}$) by dividing by 3 ($\dfrac{35 \div 5}{4 \div 2} \times \dfrac{2 \div 2}{15 \div 5} = \frac{1}{2} \times \frac{1}{3}$). No more cancellation is possible. ④ Multiply the numerators and denominators ($7 \times \frac{1}{2} \times 3 = \frac{7}{6}$). ⑤ Reduce as required ($\frac{7}{6} = 7 \div 6 = \mathbf{1\frac{1}{6}}$).

PRACTICE PROBLEMS

Dividing Mixed Numbers

1. $4\frac{2}{3} \div 3\frac{1}{2}$
2. $8\frac{1}{9} \div 6\frac{2}{3}$
3. $7\frac{1}{8} \div 5\frac{3}{4}$
4. $3\frac{3}{8} \div 2\frac{1}{2}$
5. $120\frac{2}{3} \div 2\frac{1}{2}$
6. $316\frac{1}{2} \div 2\frac{1}{2}$
7. $3\frac{1}{3} \div 3\frac{1}{3}$
8. $4\frac{1}{2} \div 2\frac{1}{4}$
9. $10\frac{1}{3} \div 3\frac{1}{9}$
10. $40\frac{5}{8} \div 3\frac{1}{5}$
11. An automatic screw machine is set up to turn $1\frac{1}{2}''$ shafts that are $\frac{1}{2}''$ in diameter. How many shafts can be turned from a $109\frac{1}{2}''$ piece of bar stock? Disregard the kerf.

12. How many $1\frac{1}{4}$ oz bottles can be filled from a beaker containing $17\frac{1}{2}$ oz?

Dividing Complex Fractions

A *complex fraction* is a fraction that has a fraction, improper fraction, mixed number, or a mathematical process in its numerator, denominator, or both. For example, $\dfrac{\frac{1}{2}}{4}$, $\dfrac{8}{\frac{1}{3}}$, $\dfrac{3\frac{1}{3}}{\frac{1}{3}}$, and $\dfrac{2 + \frac{1}{2}}{3}$ are complex fractions.

Numerator Fractions. A complex fraction with a fraction in the numerator is the same as a fraction divided by a whole number. To divide a fraction by a whole number, multiply the denominator of the numerator fraction by the denominator of the complex fraction. Then use cancellation if possible, carry the numerator, and reduce as required. See Figure 4-13.

For example, to divide ³⁄₅ by 8, multiply 5 (denominator) by 8 (whole number). Carry the 3 (³⁄₅ × 8 = ³⁄₄₀).

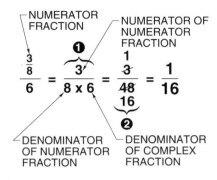

① MULTIPLY DENOMINATOR OF NUMERATOR FRACTION BY DENOMINATOR OF COMPLEX FRACTION

② REDUCE AS REQUIRED

① Multiply 8 × 6 (8 × 6 = 48) and place under the numerator 3. ② Reduce ³⁄₄₈ by dividing the numerator and denominator by 3 (³⁄₄₈ ÷ ³⁄₃ = ¹⁄₁₆).

Denominator Fractions. A complex fraction with a fraction in the denominator is the same as a whole number divided by a fraction. To divide a whole number by a fraction, change the numerator (whole number) into fraction form and invert the divisor fraction. Then use cancellation if possible, multiply the numerators and denominators, and reduce as required. See Figure 4-14.

For example, the complex fraction $\frac{9}{\frac{2}{3}}$ is the same as dividing 9 by ²⁄₃. To divide 9 by ²⁄₃, change the 9 (whole number) to ⁹⁄₁ (fraction form). Invert ²⁄₃ (divisor frac-

DIVIDING COMPLEX FRACTIONS– NUMERATOR FRACTIONS

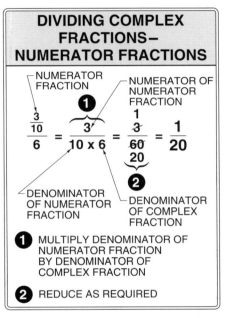

① MULTIPLY DENOMINATOR OF NUMERATOR FRACTION BY DENOMINATOR OF COMPLEX FRACTION

② REDUCE AS REQUIRED

Figure 4-13. To divide a complex fraction with a fraction in the numerator, multiply the denominator of the numerator fraction by the denominator of the complex fraction and carry the numerator.

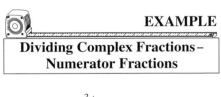

EXAMPLE

Dividing Complex Fractions– Numerator Fractions

1. Divide $\frac{\frac{3}{8}}{6}$.

tion). Multiply the numerators and denominators ($\frac{3 \times 9}{2 \times 1} = ^{27}\!/_2$). Reduce as required ($^{27}\!/_2 = 27 \div 2 = \mathbf{13\frac{1}{2}}$).

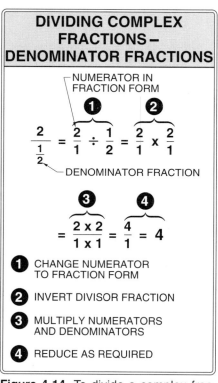

DIVIDING COMPLEX FRACTIONS – DENOMINATOR FRACTIONS

Figure 4-14. To divide a complex fraction with a fraction in the denominator, change to fraction form, invert the divisor fraction, and multiply numerators and denominators.

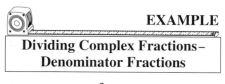

EXAMPLE

Dividing Complex Fractions – Denominator Fractions

1. Divide $\dfrac{3}{^1\!/_4}$.

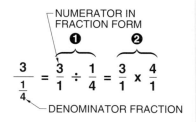

① Change the numerator to fraction form ($3 = ^3\!/_1$). ② Invert the divisor fraction ($^4\!/_1$). ③ Multiply the numerators and denominators ($\frac{3 \times 4}{1 \times 1} = ^{12}\!/_1$). ④ Reduce as required ($^{12}\!/_1 = \mathbf{12}$).

Mixed Numbers. A complex fraction with a mixed number in the numerator and denominator is the same as dividing two mixed numbers. To divide a mixed number by a mixed number, change both mixed numbers to improper fractions and invert the divisor fraction. Then use cancellation if possible, multiply the numerators and denominators, and reduce as required. See Figure 4-15.

DIVIDING COMPLEX FRACTIONS – MIXED NUMBERS

① **②**

$$\frac{2\frac{1}{2}}{1\frac{1}{3}} = \frac{5}{2} \div \frac{4}{3} = \frac{5}{2} \times \frac{3}{4}$$

MIXED NUMBERS

③ **④**

$$= \frac{5 \times 3}{2 \times 4} = \frac{15}{8} = 1\frac{7}{8}$$

① CHANGE MIXED NUMBERS TO IMPROPER FRACTIONS

② INVERT DIVISOR FRACTION

③ MULTIPLY NUMERATORS AND DENOMINATORS

④ REDUCE AS REQUIRED

Figure 4-15. To divide a complex fraction with mixed numbers in the numerator and denominator, change to improper fractions, invert the divisor fraction, and multiply numerators and denominators.

For example, to divide $1\frac{2}{3}$ by $3\frac{1}{2}$, change $1\frac{2}{3}$ and $3\frac{1}{2}$ to improper fractions ($1\frac{2}{3} = \frac{5}{3}$ and $3\frac{1}{2} = \frac{7}{2}$). Invert $\frac{7}{2}$ (divisor fraction). Multiply the numerators and denominators ($\frac{5 \times 2}{3 \times 7} = \frac{10}{21}$).

 EXAMPLES

Dividing Complex Fractions–Mixed Numbers

1. Divide $\dfrac{3\frac{1}{4}}{1\frac{1}{2}}$.

① **②**

$$\frac{3\frac{1}{4}}{1\frac{1}{2}} = \frac{13}{4} \div \frac{3}{2} = \frac{13}{4} \times \frac{2}{3}$$

MIXED NUMBERS

③ **④**

$$= \frac{13 \times 2}{4 \times 3} = \frac{26}{12} = 2\frac{1}{6}$$

① CHANGE MIXED NUMBER TO IMPROPER FRACTIONS

② INVERT DIVISOR FRACTION

③ MULTIPLY NUMERATORS AND DENOMINATORS

④ REDUCE AS REQUIRED

① Change mixed numbers to improper fractions ($3\frac{1}{4} = \frac{13}{4}$ and $1\frac{1}{2} = \frac{3}{2}$). ② Invert the divisor fraction ($\frac{2}{3}$). ③ Multiply the numerators and denominators ($\frac{13 \times 2}{4 \times 3} = \frac{26}{12}$). ④ Reduce as required ($\frac{26}{12} = 2\frac{1}{6}$).

Other Complex Fractions. To solve a complex fraction with a mathematic process(es) in the numerator and/or the denominator, solve the process(es). Divide the numerator by the denominator. See Figure 4-16.

For example, $\dfrac{\frac{5}{9} + \frac{2}{3}}{\frac{2}{3}}$ is a complex fraction with addition of fractions in the numerator. Add $\frac{5}{9}$ to $\frac{2}{3}$ ($\frac{5}{9} + \frac{2}{3} = \frac{5}{9} + \frac{6}{9} = \frac{11}{9}$).

DIVIDING COMPLEX FRACTIONS – NUMERATOR FRACTIONS

MATH PROCESS

$$\frac{(\frac{1}{3} \times \frac{3}{5}) + \frac{1}{3}}{\frac{1}{4} + \frac{3}{8}} \qquad \mathbf{1}$$

$$= \frac{\frac{3}{15} + \frac{1}{3}}{\frac{2}{8} + \frac{3}{8}} = \frac{\frac{3}{15} \times \frac{5}{15}}{\frac{5}{8}} = \frac{\frac{8}{15}}{\frac{5}{8}}$$

$$\mathbf{2} \qquad \mathbf{3}$$

$$= \frac{8}{15} \div \frac{5}{8} = \frac{8}{15} \times \frac{8}{5}$$

$$\mathbf{4}$$

$$= \frac{8 \times 8}{15 \times 5} = \frac{64}{75}$$

1 SOLVE MATH PROCESS

2 DIVIDE NUMERATOR BY DENOMINATOR

3 INVERT DIVISOR FRACTION

4 MULTIPLY NUMERATORS AND DENOMINATORS

Figure 4-16. To divide a complex fraction with mathematic processes, solve the mathematic processes, divide numerator by denominator, invert divisor fraction, and multiply numerators and denominators.

Divide $^{11}\!/_9$ by $^2\!/_3$. Invert $^2\!/_3$. Cancel 9 (of $^{11}\!/_9$) and 3 (of $^3\!/_2$) by dividing by 3 ($\frac{11}{9 \div 3} \times \frac{3 \div 3}{2} = {^{11}}\!/_3 \times$

$^1\!/_2$). No more cancellation is possible. Multiply the numerators and denominators ($\frac{11 \times 1}{3 \times 2} = {^{11}}\!/_6 = 11 \div 6 = \mathbf{1^5\!/_6}$).

Any combination of mathematic processes (addition, subtraction, multiplication, and division) can be used to form complex fractions. To solve complex fractions with addition, multiplication, or division of fractions in both the numerator and denominator, solve the math processes first and then solve the complex fraction.

EXAMPLES

Dividing Complex Fractions– Math Processes

1. Divide ($^1\!/_2 \times {^1}\!/_4$) + $^3\!/_8$ by $^3\!/_{16} + ^3\!/_4$.

MATH PROCESS

$$\frac{(\frac{1}{2} \times \frac{1}{4}) + \frac{3}{8}}{\frac{3}{16} + \frac{3}{4}}$$

$$\mathbf{1}$$

$$= \frac{\frac{1}{8} + \frac{3}{8}}{\frac{3}{16} + \frac{3}{4}} = \frac{\frac{4}{8}}{\frac{15}{16}}$$

$$\mathbf{2} \qquad \mathbf{3}$$

$$= \frac{4}{8} \div \frac{15}{16} = \frac{4}{8} \times \frac{16}{15}$$

$$= \frac{\overbrace{4 \times 16}}{8 \times 15} = \frac{\overset{\textcircled{\scriptsize 5}}{\overset{8}{\cancel{64}}}}{\underset{15}{\cancel{120}}} = \frac{8}{15}$$

❶ SOLVE MATH PROCESS
❷ DIVIDE NUMERATOR
BY DENOMINATOR
❸ INVERT DIVISOR FRACTION
❹ MULTIPLY NUMERATORS
AND DENOMINATORS
❺ REDUCE AS REQUIRED

① Solve the math process ($\frac{1}{8}$ + $\frac{\frac{2}{8} + \frac{3}{4}}{\frac{3}{16}} = \frac{\frac{4}{8}}{\frac{15}{16}}$). ② Divide the numerator by denominator ($\frac{4}{8} \div \frac{15}{16}$). ③ Invert the divisor fraction ($\frac{4}{8} \times \frac{16}{15}$). ④ Multiply numerators and denominators ($\frac{4 \times 16}{8 \times 15} = \frac{64}{120}$). ⑤ Reduce as required ($\frac{64}{120} = \frac{8}{15}$).

PRACTICE PROBLEMS

Dividing Complex Fractions

1. $\dfrac{\frac{1}{10}}{3}$

2. $\dfrac{3}{\frac{1}{2}}$

3. $\dfrac{\frac{17}{25}}{34}$

4. $\dfrac{16\frac{2}{3}}{33\frac{1}{3}}$

5. $\dfrac{30}{\frac{4}{30}}$

6. $\dfrac{16}{\frac{1}{4}}$

7. $\dfrac{\frac{13}{16}}{2}$

8. $\dfrac{2\frac{1}{4}}{\frac{5}{6}}$

9. $\dfrac{3\frac{1}{2}}{2\frac{1}{3}}$

10. $\dfrac{6\frac{2}{9}}{8\frac{2}{3}}$

11. The oats in the sack are to be equally distributed to five horses. What fractional part of the sack of oats does each horse get?

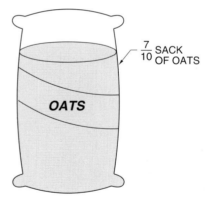

$\frac{7}{10}$ SACK OF OATS

OATS

12. A peck (pk) is $\frac{1}{4}$ of a bushel (bu). How many pecks are there in 3 bu?

1 BU   1 BU   1 BU

13. A gill (gi) is ⅛ of a quart (qt). How many gills are in 3 quarts of milk?

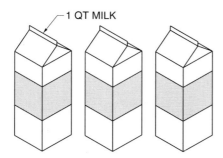

-1 QT MILK

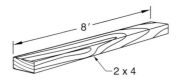

8′

2 x 4

14. Cripples (short braces) are to be cut from the 8′ 2 × 4. Disregarding the saw kerf, how many 1⅓′ cripples can be cut from the 2 × 4?

15. The remaining pie is divided into 7 equal pieces. What is the size of each piece?

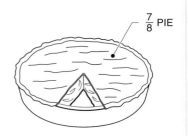

$\frac{7}{8}$ PIE

Name _____ Date _____

True-False and Completion

_____ **1.** The _____ method is a quick and easy method of multiplying a fraction by a whole number.

T   F   **2.** To multiply two fractions, multiply the numerator of one fraction by the numerator of the other fraction, and do the same with the denominators.

T   F   **3.** Fractions are divided horizontally.

T   F   **4.** More than two fractions cannot be multiplied.

_____ **5.** A(n) _____ fraction is a fraction with a mathematic process(es) in the numerator and/or the denominator.

T   F   **6.** Cancellation may be used when multiplying one improper fraction by another improper fraction.

_____ **7.** To multiply a fraction and any whole number, multiply the numerator of the fraction by the whole number and place that product over the _____.

T   F   **8.** To multiply a mixed number by a whole number, multiply the fraction of the mixed number by the whole number, multiply the whole numbers, and subtract the two products.

T   F   **9.** Vertical placement of fractions is the most common.

_____ **10.** To divide a whole number by a fraction, change the whole number into fraction form and invert the _____ fraction.

Calculations

_____ 1. $4\frac{3}{4} \div 2$

_____ 2. $6\frac{2}{3} \div 3\frac{1}{3}$

_____ 3. $\frac{5}{8} \times \frac{3}{4} \times \frac{4}{5}$

_____ 4. $4 \times \frac{3}{4} \times \frac{2}{3} \times 3\frac{1}{8}$

_____ 5. $\dfrac{7\frac{1}{2}}{2\frac{1}{4}}$

_____ 6. $\dfrac{5\frac{1}{3} \times 3\frac{3}{4}}{\frac{1}{32} \times 2\frac{1}{2}}$

_____ 7. $108\frac{3}{4} \div 10\frac{7}{8}$

_____ 8. Divide $\frac{13}{10}$ by $6\frac{1}{4}$.

_____ 9. $\frac{3}{8} \times \frac{3}{4} \times \frac{1}{2}$

_____ 10. $\dfrac{\frac{3}{4}}{12}$

_____ 11. A person owns $\frac{3}{5}$ of a building worth \$60,000.00. If $\frac{1}{4}$ of that share is sold, find the value of the part the person retains.

_____ 12. River A is $416\frac{2}{3}$ mi long. River B is $\frac{3}{5}$ of River A. River B is how long?

_____ 13. If a cubic foot of water weighs $62\frac{1}{2}$ lb and steel is $7\frac{4}{5}$ times as heavy, how much does a cubic foot of steel weigh?

_____ 14. The metal lining of a tank weighs $3\frac{3}{5}$ lb per square foot. Find how many pounds are required to line a tank with an inside surface of $237\frac{1}{2}$ sq ft.

Name _____ Date _____

1. $\frac{3}{4} \times \frac{16}{3} \times \frac{18}{24} \times \frac{12}{9}$

2. $\frac{7}{32} \times 16 \times 3\frac{1}{4}$

3. Divide $\frac{13}{10}$ by $6\frac{1}{4}$.

4. Divide 39 by $\frac{3}{2}$.

5. $\dfrac{4}{\frac{1}{2}} \times \dfrac{\frac{1}{3}}{\frac{1}{2} \times \frac{4}{2}}$

6. $2\frac{6}{10} \times \dfrac{3}{\frac{2}{3}} \times \frac{6}{8}$

7. $24 \div \dfrac{\frac{2}{3}}{2}$

8. $5\frac{5}{8} + \dfrac{\frac{3}{4} \div \frac{1}{2}}{2 + \frac{2}{3}} \div \frac{3}{4}$

9. $\frac{3}{4} \times \frac{2}{3}$

10. $\dfrac{\frac{2}{5} \times \frac{5}{6}}{\frac{2}{9} \times 4\frac{1}{2}}$

11. $\dfrac{11\frac{3}{7}}{\frac{4}{7}}$

12. If a young man is $18\frac{1}{3}$ years old and his father has lived $2\frac{1}{2}$ times as long, what is his father's age?

13. A railroad rail weighs 120 lb per foot. How much does $\frac{1}{2}'$ weigh?

14. Multiply $(3\frac{1}{2} + 4\frac{1}{4}) + 3$ by 4.

105

_____ **15.** John's house is $1\frac{3}{4}$ mi from Sarah's house. Ted lives halfway between John and Sarah. How far is John's house from Ted's house?

_____ **16.** What is $\frac{1}{3}$ of $\frac{2}{3}$?

_____ **17.** A carpenter's nail gun is $\frac{4}{5}$ full. While nailing stud walls, $\frac{1}{3}$ of the nails are used. What fractional part of the fully-loaded nail gun is used?

_____ **18.** A flooring contractor has 360 tiles to lay. After 3 hours, $\frac{3}{4}$ of the tiles are laid. How many tiles are left to lay?

_____ **19.** A cabinetmaker has 48 door pulls and uses $\frac{3}{4}$ of them on one job. How many door pulls were used on the job?

_____ **20.** Dowels for faceplates are purchased in 100 lot bags. The faceplates for three cabinets require $\frac{3}{4}$ of 1 bag. How many dowels are required for the three cabinets?

_____ **21.** How many $6\frac{1}{2}$ oz glasses can be filled from a pitcher containing 104 oz?

_____ **22.** Divide $(\frac{1}{4} \times \frac{1}{2}) + \frac{7}{8}$ by $\frac{3}{4}$.

_____ **23.** Multiply $(\frac{3}{8} + \frac{3}{4}) + 4$ by 3.

DECIMALS

A decimal is a number expressed in base 10. The two types of decimal numbers are proper decimal numbers and mixed decimal numbers. Decimals can be added, subtracted, multiplied, and divided. Fractions can be changed to decimals and decimals can be changed to fractions.

DECIMALS

A *decimal fraction* is a fraction with the denominator 10, 100, 1000, etc. The number 1 is the smallest whole number. Anything smaller than 1 is a decimal and can be divided into any number of parts. For example, .75 shows that the whole number 1 is divided into 100 parts, and 75 parts are present.

Any fraction with 10, 100, 1000, or another multiple of 10 for the denominator may be written as a decimal. For example, $\frac{3}{10}$ is .3, $\frac{35}{100}$ is .35, and $\frac{357}{1000}$ is .357 in the decimal system.

A *decimal* is a number expressed in base 10. The two types of decimal numbers are proper decimal numbers and mixed decimals. A *proper decimal number* is a decimal number that has no whole numbers.

For example, .375 is a proper decimal number. A *mixed decimal number* is a decimal number that has a whole number and a decimal number separated by a decimal point. For example, 3.14 is a mixed decimal.

A *decimal point* is the period at the left of a proper decimal number or the period that separates the parts of a mixed decimal. All numbers to the left of the decimal point are whole numbers. All numbers to the right of the decimal point are less than whole numbers. See Figure 5-1.

When reading decimal numbers or writing decimal numbers as word statements, the word "point" or "and" is used at the decimal point. Usually the last decimal number is followed by the place that it occupies. For example, 9.15 is read "nine and fifteen hundredths" or "nine point fifteen."

Sometimes the decimal numbers are stated individually. For example, 8.29 can be read "eight point two nine."

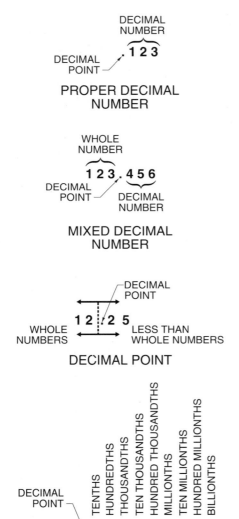

Figure 5-1. A decimal is a number expressed in base 10.

The United States monetary system is based on decimals. See Figure 5-2. The dollar ($1.00) is valued at 100 cents (100¢). Each penny is $\frac{1}{100}$ of a dollar ($.01 or 1¢). Each nickel is $\frac{5}{100}$ of a dollar ($.05 or 5¢). Each dime is $\frac{10}{100}$ of a dollar ($.10 or 10¢). Each quarter is $\frac{25}{100}$ of a dollar ($.25 or 25¢). Each half-dollar is $\frac{50}{100}$ of a dollar ($.50 or 50¢).

In measurements, more places in a decimal number indicate a higher degree of accuracy. When measuring tolerances on a machined part, the tolerance may be measured in hundredths or thousandths.

For example, two parts may have measurements of 1.250″ ±.005″ and 1.25″ ±.01″. While the overall size of the parts is the same, the degree of accuracy is more critical for the part measured in thousandths.

A *repeating decimal* is a decimal number with a repetend. A *repetend* is a group of figures of a decimal number that are repeated infinitely. The repetend of a repeating decimal is indicated with a rule above it.

For example, the fraction $\frac{1}{3}$ has a decimal equivalent of .33$\overline{33}$. The rule above the last two figures indicates that 33 (repetend) repeats infinitely (forever). Depending on the degree of accuracy required, repeating decimals are carried out to a certain number of decimal places.

U.S. MONETARY SYSTEM			
CURRENCY		VALUE	DECIMAL	FRACTION
DOLLAR	◯	**$1.00**	**1.00**	$\frac{100}{100}$
HALF-DOLLAR	◯	**$.50**	**.50**	$\frac{50}{100}$
QUARTER	◯	**$.25**	**.25**	$\frac{25}{100}$
DIME	◯	**$.10**	**.10**	$\frac{10}{100}$
NICKEL	◯	**$.05**	**.05**	$\frac{5}{100}$
PENNY	◯	**$.01**	**.01**	$\frac{1}{100}$

Figure 5-2. The United States monetary system is based on decimals.

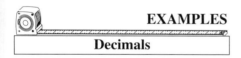

EXAMPLES

Decimals

1. Write .68 as a word statement.

The last decimal figure (8) occupies the hundredths place. The decimal .68 is written as **sixty-eight hundredths, point sixty-eight**, or **point six eight**.

2. Write 62.292 as a word statement.

The last decimal figure (2) occupies the thousandths place. The mixed decimal 62.292 is written as **sixty-two and two hundred ninety-two thousandths, sixty-two point two hundred ninety-two**, or **sixty-two point two nine two**.

3. Write thirteen hundredths as a decimal number.

The last figure of 13 (3) occupies the hundredths place. Thirteen hundredths is written as **.13**.

4. Write two hundred seven thousand, eighty-three ten millionths as a decimal number.

Write the entire number 207,083 and then place the decimal point so that the last figure (3) is in the ten millionths place. The seventh place to the right of the decimal point is the ten millionths place. Place a zero before the number in order to get a seventh place to the right of the decimal point. Two hundred seven thousand, eighty-three ten millionths is written as **.0207083**.

PRACTICE PROBLEMS

Decimals

Write as word statements.
1. .7
2. .0091
3. 6.31

Write as decimal numbers.
4. four and seven tenths
5. seventy-five hundredths
6. one hundred and forty-four thousandths
7. one hundred forty-four thousandths

Adding and Subtracting Decimals

To add or subtract decimals, align the numbers by the decimal points so columns are vertically aligned. See Figure 5-3.

Units are added to or subtracted from units, tenths to tenths, hundredths to hundredths, etc. Add or subtract same as whole numbers and place the decimal point of the sum or difference directly below the other decimal points.

For example, to add 27.08 and 9.127, align the numbers vertically by the decimal points. Add column by column (27.08 + 9.127 = 36.207). To subtract 4.15 from 10.56, align the numbers vertically by the decimal points. Subtract column by column (10.56 − 4.15 = **6.41**).

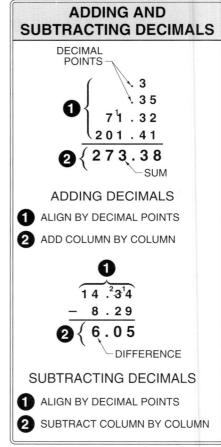

Figure 5-3. To add or subtract decimals, align by decimal points and perform the math process.

EXAMPLES

Adding and Subtracting Decimals

1. Add 27.072 and 8.923.

DECIMAL
POINTS

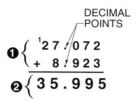

❶ ALIGN BY DECIMAL POINTS
❷ ADD COLUMN BY COLUMN

① Align the numbers vertically by the decimal points. ② Add column by column (27.072 + 8.923 = **35.995**).

2. Add .0004, 24.345, 740.01, and 1.2345.

$$\mathbf{❶}\begin{cases} \ \ \ \ .0004 \\ \ \ 24.345 \\ 740.01 \\ \ \ \ \ 1.2345 \end{cases}$$
$$\mathbf{❷}\big\{ \ 765.5899$$

❶ ALIGN BY DECIMAL POINTS
❷ ADD COLUMN BY COLUMN

① Align the numbers vertically by the decimal points. ② Add column by column (.0004 + 24.345 + 740.01 + 1.2345 = **765.5899**).

3. Subtract 275.0005 from 3000.024.

ANNEX ZERO
FOR SUBTRACTION

$$\mathbf{❶}\big\{ \begin{array}{l} {}^{2}3{}^{9}0{}^{9}0{}^{1}0.02{}^{3}4{}^{1}0 \\ -\ \ 275.0005 \end{array}$$
$$\mathbf{❷}\big\{ 2725.0235$$

❶ ALIGN BY DECIMAL POINTS
❷ ADD COLUMN BY COLUMN

① Align the numbers vertically by the decimal points and annex a zero to the ten thousandths decimal place. ② Subtract column by column (3000.024 − 275.0005 = **2725.0235**).

4. Subtract .100203 from .30405.

$$\mathbf{❶}\big\{ \begin{array}{l} .30{}^{3}4{}^{1}0{}^{4}5{}^{1}0 \\ -.100203 \end{array}$$
$$\mathbf{❷}\big\{ .203847$$

❶ ALIGN BY DECIMAL POINTS
❷ ADD COLUMN BY COLUMN

① Align the numbers vertically by the decimal points and annex a zero to the millionths decimal place. ② Subtract column by column (.30405 − .100203 = **.203847**).

PRACTICE PROBLEMS

Adding and Subtracting Decimals

1. 34.51 + 94.3545 + 2.09847
2. 30,234.357 − 345.984
3. 45.943 + 9.4329
4. 474.84005 − 89.459323
5. 94.097 − .00875
6. 388.987 + 474.00653 + .93474 + .000409
7. 940.0098 − .9929
8. A cabinetmaking project requires 35.875 board feet (bd ft) of cherry, 129.125 bd ft of pine, and 16.125 bd ft of fir.

What is the total amount of board feet required?

9. The total wall area of a room (including door and window openings) is 480 sq ft. The door opening is 16.675 sq ft. Two window openings total 20.375 sq ft. What is the approximate area of wall to be painted?

10. How much metal is machined off the Wheel?

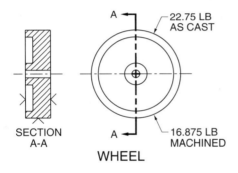

SECTION A-A

WHEEL

Multiplying Decimals

To multiply decimals, multiply same as whole numbers. Then from the right of the product, point off toward the left the number of decimal places equal to the number of decimal places in the quantities multiplied. See Figure 5-4. For example, to multiply 20.45 by 3.15, multiply same as whole numbers and point off four places from the right (20.45 × 3.15 = **64.4175**).

MULTIPLYING DECIMALS

MULTIPLICAND

❶ $\begin{cases} \overset{2}{2} . \overset{4}{3} 6 \\ \times \ . 1 7 \\ \hline \overset{1}{1} \overset{1}{6} 5 2 \\ 2 3 6 0 \end{cases}$ ←MULTIPLIER

❷ $\{ . 4 0 1 2$

❶ MULTIPLY SAME AS WHOLE NUMBER

❷ POINT OFF EQUAL TO DECIMAL PLACES IN MULTIPLICAND AND MULTIPLIER

MOVE DECIMAL POINT TO RIGHT AS MANY PLACES AS ZEROS IN MULTIPLIER

37.954 x 100 = 37.95.4

MULTIPLYING DECIMALS BY 10,100,1000, ETC.

MULTIPLY MULTIPLICAND BY 100 AND DIVIDE BY 2

$.4 \times 50 = \dfrac{.4 \times 100}{2} = 20$

MULTIPLYING DECIMALS BY 50

MULTIPLY MULTIPLICAND BY 100 AND DIVIDE BY 4

$.4 \times 25 = \dfrac{.4 \times 100}{4} = 10$

MULTIPLYING DECIMALS BY 25

Figure 5-4. Multiply decimals same as whole numbers and point off the number of decimal places in the multiplicand and multiplier.

To multiply a decimal or a mixed decimal by 10, 100, 1000, etc., move the decimal point of the multiplicand as many places to the right as there are zeros in the multiplier. If there are not enough figures, annex zeros to the right. For example, to multiply 37.954 by 100, move the decimal point two places to the right (37.954 × 100 = **3795.4**).

To multiply a decimal or a mixed decimal by 50, multiply the multiplicand by 100 and divide by 2. To multiply a decimal or a mixed decimal by 25, multiply the multiplicand by 100 and divide by 4.

EXAMPLES

Multiplying Decimals

1. Multiply .875 by .37.

❶ $\begin{cases} \overset{2}{.}\overset{4}{8}\overset{1}{7}5 & \text{—MULTIPLICAND} \\ \text{x} \quad .37 & \text{—MULTIPLIER} \\ \overline{6125} \\ {}^{1}26250 \end{cases}$

❷ $\left\{ .32375 \right.$

❶ MULTIPLY SAME AS WHOLE NUMBERS
❷ POINT OFF EQUAL TO DECIMAL PLACES IN MULTIPLICAND AND MULTIPLIER

① Multiply same as whole numbers (875 × 37 = 32,375). ② Point off five decimal places from right (.875 × .37 = **.32375**).

2. Multiply 4.7023 by 1.092.

❶ $\begin{cases} \overset{6}{\overset{1}{4}}.7\overset{2}{0}\overset{2}{2}3 \\ \text{x} \quad 1.092 \\ \overline{94046} \\ 4^{1}23207 \\ {}^{1}47023 \end{cases}$

❷ $\left\{ 5.1349116 \right.$

❶ MULTIPLY SAME AS WHOLE NUMBERS
❷ POINT OFF EQUAL TO DECIMAL PLACES IN MULTIPLICAND AND MULTIPLIER

① Multiply same as whole numbers (47,023 × 1092 = 51,349,116). ② Point off seven decimal places from right (4.7023 × 1.092 = **5.1349116**).

3. Multiply .00123 by .01023.

❶ $\begin{cases} .00123 \\ \text{x} .01023 \\ \overline{00^{1}369} \\ 00246 \\ 00123 \end{cases}$

❷ $\left\{ .0000125829 \right.$

❶ MULTIPLY SAME AS WHOLE NUMBERS
❷ POINT OFF EQUAL TO DECIMAL PLACES IN MULTIPLICAND AND MULTIPLIER

① Multiply same as whole numbers (123 × 1023 = 125,829). ② Point off ten decimal places from right (.00123 × .01023 = **.0000125829**). It is necessary to annex four zeros to obtain ten places.

4. Multiply .4505 by 1000.

❶
.4505 x 1000 = 450.5

❶ MOVE DECIMAL POINT
  THREE PLACES TO RIGHT

① Multiply .4505 by 1000 by moving the decimal point three places to the right (.4505 × 1000 = **450.5**).

5. Multiply .350 by 50.

$$❶\left\{\frac{.350 \times 100}{2} = 17.5\right.$$

❶ MULTIPLY MULTIPLICAND BY 100
  AND DIVIDE BY 2

① Multiply .350 by 50 by multiplying the multiplicand by 100 and dividing by 2 (.350 × 100 ÷ 2 = **17.5**).

6. Multiply .350 by 25.

$$❶\left\{\frac{.350 \times 100}{4} = 8.75\right.$$

❶ MULTIPLY MULTIPLICAND BY 100
  AND DIVIDE BY 4

① Multiply .350 by 25 by multiplying the multiplicand by 100 and dividing by 4 (.350 × 100 ÷ 4 = **8.75**).

PRACTICE PROBLEMS

Multiplying Decimals

1. .35 × 4
2. .785 × 25 *(3 places)*
3. .287 × .356 *(6 places)*
4. .002 × .014 *(6 places)*

5. 1.0034 × 2.503 *(7 places)*
6. .35 × 12.5 *(3 places)*
7. 56.98 × 1000
8. A wire fence consists of 16.5 sections each 9.75′ long. How long is the fence? *(3 places)*
9. What is the total rise of the stairs?

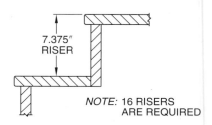

7.375″
RISER

NOTE: 16 RISERS
ARE REQUIRED

STAIR DETAIL

10. To complete a job, a plumber needs 10 sections of pipe that are each 24.125″ in length. How much pipe is needed?
11. 24.575 × 50
12. 2.25 × 25

Dividing Decimals

To divide decimals, divide same as whole numbers. Then point off from right to left as many decimal places as the difference between the number of decimal places in the dividend and in the divisor. See Figure 5-5.

If the dividend has fewer decimal places than the divisor, annex zeros to the dividend. There must be at least as many decimal places in the dividend as in the divisor.

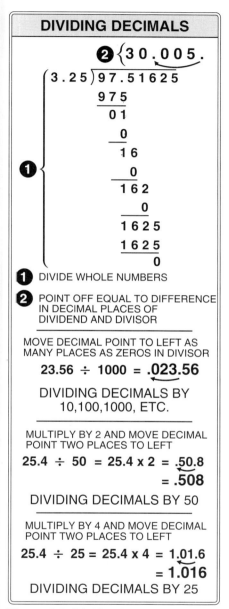

DIVIDING DECIMALS

❷ {30.005.

3.25)97.51625

975

01

0

16

❶ 0

162

0

1625

1625

0

❶ DIVIDE WHOLE NUMBERS

❷ POINT OFF EQUAL TO DIFFERENCE IN DECIMAL PLACES OF DIVIDEND AND DIVISOR

MOVE DECIMAL POINT TO LEFT AS MANY PLACES AS ZEROS IN DIVISOR

23.56 ÷ 1000 = .023.56

DIVIDING DECIMALS BY 10,100,1000, ETC.

MULTIPLY BY 2 AND MOVE DECIMAL POINT TWO PLACES TO LEFT

25.4 ÷ 50 = 25.4 x 2 = .50.8

= .508

DIVIDING DECIMALS BY 50

MULTIPLY BY 4 AND MOVE DECIMAL POINT TWO PLACES TO LEFT

25.4 ÷ 25 = 25.4 x 4 = 1.01.6

= 1.016

DIVIDING DECIMALS BY 25

Figure 5-5. Divide decimals same as whole numbers. Point off the number of decimal places equal to the difference between the decimal places in the dividend and the divisor.

There may be as many zeros annexed as needed because adding zeros to the right of a decimal does not alter its value.

For example, to divide 16.75 by 2.5, divide the same as whole numbers ($1675 ÷ 25 = 67$). Point off one decimal place (difference in number of decimal places) from right to left ($16.75 ÷ 2.5 = $ **6.7**).

To divide a decimal or mixed decimal by 10, 100, 1000, etc., move the decimal point one place to the left for each zero in the divisor. Annex zeros as needed.

To divide a decimal or mixed decimal by 50, multiply by 2 and move the decimal point two places to the left. Annex zeros as needed.

To divide a decimal or mixed decimal by 25, multiply by 4 and move the decimal point two places to the left. Annex zeros as needed.

To divide a decimal or mixed decimal by 12.5, multiply by 8 and move the decimal point two places to the left. Annex zeros as needed.

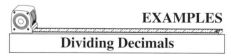

EXAMPLES

Dividing Decimals

1. Divide 37.24588 by 124.

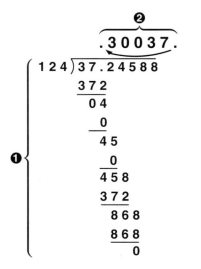

❶ DIVIDE SAME AS WHOLE NUMBERS
❷ POINT OFF DECIMAL PLACES EQUAL
TO DIFFERENCE IN DECIMAL PLACES
OF DIVIDEND AND DIVISOR

① Divide same as whole numbers ($3{,}724{,}588 \div 124 = 30{,}037$). ② Point off five places to the left ($37.24588 \div 124 = \mathbf{.30037}$).

2. Divide .450 by 1000.

❶{ $.450 \div 1000 = .\underset{\curvearrowleft}{000}.450$

❶ MOVE DECIMAL POINT
THREE PLACES TO LEFT

① Divide .450 by 1000 by moving the decimal point three places to the left ($.450 \div 1000 = \mathbf{.000450}$).

3. Divide .475 by 50.

❶
$47.5 \div 50 = 47.5 \times 2 = .\underset{\curvearrowleft}{95}. = \mathbf{.95}$

❶ MOVE DECIMAL POINT
TWO PLACES TO LEFT

① Divide 47.5 by 50 by multiplying by 2 and moving the decimal point two places to the left ($47.5 \times 2 = 95 = \mathbf{.95}$).

4. Divide .876447 by 27.

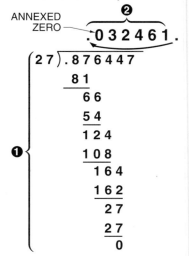

ANNEXED ZERO

❶ DIVIDE SAME AS WHOLE NUMBERS
❷ POINT OFF DECIMAL PLACES EQUAL
TO DIFFERENCE IN DECIMAL PLACES
OF DIVIDEND AND DIVISOR

① Divide same as whole numbers ($876{,}447 \div 27 = 32{,}461$). ② Point off six places to the left and annex one zero ($.876447 \div 27 = \mathbf{.032461}$).

PRACTICE PROBLEMS

Dividing Decimals

1. $3.036 \div .06$
2. $3.728 \div .16$
3. $.864 \div .024$
4. $10.044 \div .36$

5. 12 ÷ .7854 *(7 places)*
6. 2.34 ÷ .211 *(8 places)*
7. .125 ÷ 8000 *(9 places)*
8. A line 4.5″ long is divided into six equal parts. How long is each part?
9. 3.16 ÷ 10
10. 40.5 ÷ 25
11. What is the length of C on the Step Gauge? All steps are equally spaced. *(3 places)*

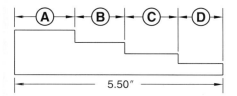

STEP GAUGE

12. Find the center-to-center distance between holes F and G of the Hole Gauge. *(4 places)*

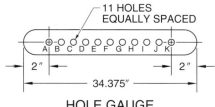

HOLE GAUGE

13. 34.75 ÷ 1000 *(5 places)*
14. 245.25 ÷ 50 *(3 places)*
15. 45.75 ÷ 25

Changing Fractions to Decimals

To change a fraction to a decimal, divide the numerator by the denominator.

Annex zeros if needed. See Figure 5-6. For example, to change $\frac{5}{16}$ to a decimal number, divide 5 by 16 ($\frac{5}{16}$ = 5 ÷ 16 = **.3125**). Tables also show the equivalent decimals for fractions. See Appendix.

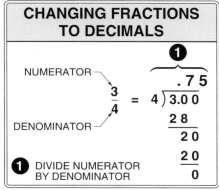

Figure 5-6. To change a fraction to a decimal, divide the numerator by denominator.

EXAMPLES

Changing Fractions to Decimals

1. Change $\frac{5}{11}$ to a decimal.

① Divide 5 by 11 (5 ÷ 11 = **.45̄4̄5̄**).

2. Change $\frac{7}{64}$ to a decimal.

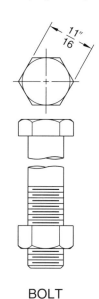

NUMERATOR ❶

$$\frac{7}{64} = \begin{array}{r} .1\,0\,9\,3\,7\,5 \\ 6\,4\overline{\smash{)}7.0\,0\,0\,0\,0\,0} \\ \underline{6\,4} \\ 6\,0 \\ \underline{0} \\ 6\,0\,0 \\ \underline{5\,7\,6} \\ 2\,4\,0 \\ \underline{1\,9\,2} \\ 4\,8\,0 \\ \underline{4\,4\,8} \\ 3\,2\,0 \\ \underline{3\,2\,0} \\ 0 \end{array}$$

DENOMINATOR

❶ DIVIDE NUMERATOR BY DENOMINATOR

① Divide 7 by 64 (7 ÷ 64 = **.109375**).

PRACTICE PROBLEMS

Changing Fractions to Decimals

Change the following fractions to decimals.

1. $\frac{5}{32}$ *(5 places)*

2. $\frac{35}{100}$

3. $\frac{47}{150}$ *(4 places)*

4. $\frac{17}{32}$ *(5 places)*

5. $\frac{3}{8}$ *(3 places)*

6. $\frac{9}{32}$ *(5 places)*

7. $\frac{3}{4}$

8. $\frac{19}{64}$ *(6 places)*

9. The distance across flats on a bolt head is $\frac{11}{16}''$. What is the decimal distance across flats on the Bolt? *(4 places)*

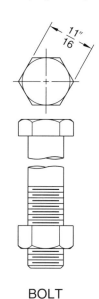

BOLT

10. What is the missing dimension X (in decimals) on the Clamp?

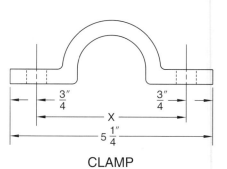

CLAMP

Changing Decimals to Fractions

To change a decimal to a fraction, use the figures as the numerator. For the denominator, place a 1 followed by as many zeros as there are figures to the right of the decimal point in the quantity. See Figure 5-7.

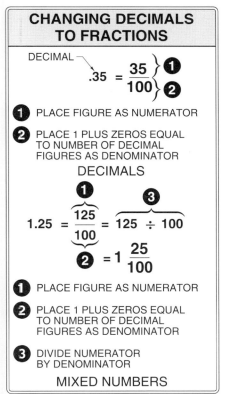

CHANGING DECIMALS TO FRACTIONS

DECIMAL

$$.35 = \frac{35}{100} \left. \begin{array}{l} \text{❶} \\ \text{❷} \end{array} \right.$$

❶ PLACE FIGURE AS NUMERATOR

❷ PLACE 1 PLUS ZEROS EQUAL TO NUMBER OF DECIMAL FIGURES AS DENOMINATOR

DECIMALS

$$1.25 = \frac{\overset{❶}{\overbrace{125}}}{\underset{❷}{100}} = \overset{❸}{\overbrace{125 \div 100}}$$

$$\overset{❷}{=} 1 \frac{25}{100}$$

❶ PLACE FIGURE AS NUMERATOR

❷ PLACE 1 PLUS ZEROS EQUAL TO NUMBER OF DECIMAL FIGURES AS DENOMINATOR

❸ DIVIDE NUMERATOR BY DENOMINATOR

MIXED NUMBERS

Figure 5-7. Decimals and mixed decimals can be changed to fractions.

For example, to change the decimal .71 to a fraction, use the 71 as the numerator and place a 1 followed by two zeros (number of figures to right of the decimal point) as the denominator (.71 = $^{71}/_{100}$).

To change a mixed decimal to a fraction, drop the decimal point and use the figure as the numerator. For the denominator, place a 1 followed by as many zeros as there are figures to the right of the decimal point in the quantity.

For example, to change 6.22 to a fraction, drop the decimal point (622). Use 622 as the numerator and place a 1 followed by two zeros (number of figures to the right of the decimal point) as the denominator ($^{622}/_{100}$). Reduce $^{622}/_{100}$ to 6 $^{22}/_{100}$, which further reduces to $6^{11}/_{50}$ (6.22 = $6^{11}/_{50}$).

EXAMPLES

Changing Decimals to Fractions

1. Change .3 to a fraction.

DECIMAL

$$.3 = \frac{3}{10} \left. \begin{array}{l} \text{❶} \\ \text{❷} \end{array} \right.$$

❶ PLACE FIGURE AS NUMERATOR

❷ PLACE 1 PLUS ZEROS EQUAL TO NUMBER OF DECIMAL FIGURES AS DENOMINATOR

① Use the 3 as the numerator. ② Place 1 followed by a zero as the denominator (.3 = $^{3}/_{10}$).

2. Change .045 to a fraction.

$$.045 = \frac{\overbrace{045}}{\underbrace{1000}} = \frac{45}{1000}$$

❶ PLACE FIGURE AS NUMERATOR
❷ PLACE 1 PLUS ZEROS EQUAL
   TO NUMBER OF DECIMAL
   FIGURES AS DENOMINATOR

① Use the 045 as the numerator. ② Place 1 followed by three zeros as the denominator ($^{045}/_{1000}$). In this example, the zero in the numerator has no value and can be dropped ($^{045}/_{1000} = {}^{45}/_{1000}$).

3. Change the decimal 7.002 to a fraction.

$$7.002 = \frac{\overbrace{7002}}{\underbrace{1000}} = \overbrace{7002 \div 1000}$$

$$= 7\frac{2}{1000} = 7\frac{1}{500}$$

❶ PLACE FIGURE AS NUMERATOR
❷ PLACE 1 PLUS ZEROS EQUAL
   TO NUMBER OF DECIMAL
   FIGURES AS DENOMINATOR
❸ DIVIDE NUMERATOR
   BY DENOMINATOR

① Use 7002 as the numerator. ② Place a 1 followed by three zeros as the denominator ($^{7002}/_{1000}$).

③ Reduce $^{7002}/_{1000}$ to $7^{2}/_{1000}$, which further reduces to $7^{1}/_{500}$ (7.002 = $7^{2}/_{1000} = \mathbf{7^{1}/_{500}}$).

PRACTICE PROBLEMS

Changing Decimals to Fractions

Change the following decimals to fractions.

1. .325
2. .004
3. .0205
4. 9.3
5. .1930
6. 7.114
7. 8.375
8. Change pi (3.14) to a fraction.
9. The specific gravity of aluminum is 2.56. What is the specific gravity as a fraction?
10. A machined part has a dimensional tolerance of ±.06. What is the tolerance as a fraction?
11. A #32 drill has a diameter of .116″. What is the diameter as a fraction?
12. A transmission chain has a roller diameter of .469″. What is the diameter as a fraction?
13. On the Bracket, what are the overall length, height, and width as fractions?

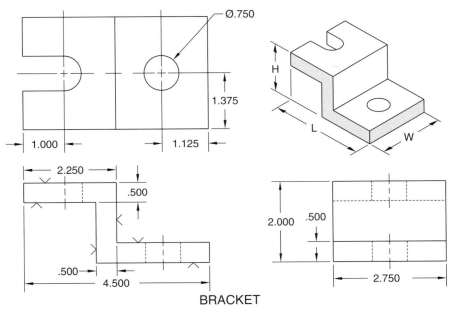

BRACKET

Changing Decimals to Fractions with a Given Denominator

To change a decimal to a fraction with a given denominator, change the decimal to a fraction and multiply the numerator and denominator by the given denominator. See Figure 5-8.

For example, to change .245 to twelfths, change .245 to a fraction (.245 = $^{245}/_{1000}$). Multiply 245 and 1000 by 12 ($\frac{245 \times 12}{1000 \times 12}$ = $^{2940}/_{12,000}$ = $^{2.94}/_{12}$).

The approximate value is $^3/_{12}$.

CHANGING DECIMALS TO FRACTIONS WITH GIVEN DENOMINATORS

GIVEN DENOMINATOR ❶ ❷

$$.475 = \frac{}{3} = \frac{475 \times 3}{1000 \times 3} = \frac{1425}{3000}$$

DECIMAL ❸ APPROXIMATE VALUE

$$= \frac{1.425}{3} = \frac{1}{3}$$

❶ MULTIPLY NUMERATOR AND DENOMINATOR OF DECIMAL IN FRACTION FORM BY GIVEN DENOMINATOR

❷ REDUCE FRACTION

❸ ROUND TO NEAREST FRACTION OF GIVEN DENOMINATOR

Figure 5-8. Change the decimal to a fraction and multiply the numerator and denominator by the given denominator.

EXAMPLES

Changing Decimals to Fractions with a Given Denominator

1. Change .564 to 64ths.

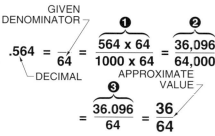

GIVEN DENOMINATOR

❶ MULTIPLY NUMERATOR AND DENOMINATOR OF DECIMAL IN FRACTION FORM BY GIVEN DENOMINATOR

❷ REDUCE FRACTION

❸ ROUND TO NEAREST FRACTION OF GIVEN DENOMINATOR

Change .564 to a fraction (.564 = $^{564}/_{1000}$). ① Multiply 564 and 1000 by 64 ($\frac{564 \times 64}{1000 \times 64}$ = $^{36,096}/_{64,000}$). ② Reduce the fraction ($^{36,096}/_{64,000}$ = $^{36.096}/_{64}$). ③ Round to the nearest fraction of the given denominator. The approximate value is $^{36}/_{64}$.

2. Change 1.33 to 9ths.

$$1.33 \; = \; \overset{❶}{\overbrace{\frac{133 \times 9}{100 \times 9}}} \; = \; \overset{❷}{\overbrace{\frac{1197}{900}}}$$

$$= \; \overset{❸}{\overbrace{\frac{11.97}{9}}} \; = \; \frac{12}{9} \; = \; 1\frac{3}{9}$$

❶ MULTIPLY NUMERATOR AND DENOMINATOR OF DECIMAL IN FRACTION FORM BY GIVEN DENOMINATOR

❷ REDUCE FRACTION

❸ ROUND TO NEAREST FRACTION OF GIVEN DENOMINATOR

Change 1.33 to a fraction (1.33 = $^{133}/_{100}$). ① Multiply 133 and 100 by 9 ($\frac{133 \times 9}{100 \times 9}$ = $^{1197}/_{900}$). ② Reduce the fraction ($^{1197}/_{900}$ = $^{11.97}/_{9}$). ③ Round to the nearest fraction of the given denominator ($^{12}/_{9}$ = 1$^{3}/_{9}$). The approximate value is **1$^{3}/_{9}$**.

PRACTICE PROBLEMS

Changing Decimals to Fractions with a Given Denominator

1. Change .756 to 12ths.
2. Change .875 to 64ths.
3. Change .719 to 32nds.
4. Change .45 to 16ths.
5. Change .622 to 8ths.

Name _____ Date _____

True-False and Completion

T   F   **1.** The decimal .350 equals $^{35}/_{1000}$.

_____   **2.** A(n) _____ decimal number is a decimal number that has no whole numbers.

T   F   **3.** To change a fraction to a decimal number, divide the denominator by the numerator.

_____   **4.** A(n) _____ decimal number is a decimal number that has a whole number and a decimal number separated by a decimal point.

_____   **5.** In measurements, more places in a decimal number indicate a higher degree of _____ .

T   F   **6.** A decimal is a number expressed in base 10.

_____   **7.** A(n) _____ is a group of figures of a decimal number that are repeated infinitely.

T   F   **8.** The number 4.89 is a proper decimal number.

T   F   **9.** The U.S. monetary system is based on the decimal system.

_____   **10.** A(n) _____ is the period at the left of a proper decimal number or that separates the parts of a mixed decimal.

Calculations

Change the following to fractions and reduce to lowest terms.

_____   **1.** 25.25

_____   **2.** 836.125

_____   **3.** .250

_____ **4.** 4.375

_____ **5.** 14.75

_____ **6.** A machinist cuts five pieces from a bar of steel 25.5′ long. How much bar is left if the five pieces are 3.5′, 3.75′, 4.75′, 4.24′, and 4.0625′ long? Disregard the saw kerf. *(4 places)*

_____ **7.** Change .9375 to 16ths.

_____ **8.** .05643 ÷ .00456 *(3 places)*

_____ **9.** 9.4565 − 3.094562 *(6 places)*

_____ **10.** 3.54 × 25

_____ **11.** 12 ÷ .54

_____ **12.** .30084 + 345.323 + 87.0038 + .080926 *(6 places)*

_____ **13.** 134.124 × 1000

_____ **14.** An iron bar is 24″ long, 2 ⅝″ wide, and ¾″ thick. A cubic inch of iron weighs .261 lb. (cu in. = l × w × t). What is the weight of the iron bar? *(5 places)*

_____ **15.** Change .625 to 8ths.

_____ **16.** 94.094 − 38.003493 *(6 places)*

_____ **17.** .0454 + .857 + 67.0989 + .00004487 + 5.97 *(8 places)*

_____ **18.** .058 + .2591 + 2.15 *(4 places)*

Change the following to decimal numbers.

_____ **19.** $6\frac{7}{8}$ *(3 places)*

_____ **20.** $8\frac{5}{8}$ *(3 places)*

_____ **21.** $12\frac{9}{16}$ *(4 places)*

_____ **22.** $4\frac{5}{32}$ *(5 places)*

_____ **23.** $829\frac{19}{32}$ *(5 places)*

_____ **24.** A pound of white brass is composed of .64 lb of tin, .02 lb of copper, and .34 lb of zinc. How many pounds of tin are in a 7.125 lb white brass casting?

_____ **25.** $1\frac{7}{8} \times .32$

_____ **26.** $4\frac{3}{5} \div .8$

_____ **27.** Change 4.33 to 3rds.

_____ **28.** .453 ÷ .3

_____ **29.** Change three hundred twenty-two ten thousandths to a decimal.

_____ **30.** What length copper pipe is required so that two 3.75′ long pieces, four 2.875′ long pieces, and one 12.75′ long piece can be cut from it?

_____ **31.** Change 5.802 to a word statement.

_____ **32.** Change seventeen and twenty-six hundredths to a decimal number.

_____ **33.** A contractor seals four driveways with areas of 84.8 sq yd, 124.1 sq yd, 98.9 sq yd, and 344.7 sq yd. How many drums of sealant will be needed if each drum covers 150 sq yd?

_____ **34.** Change $9\frac{3}{5}$ to a decimal number.

_____ **35.** Subtract forty-four ten thousandths from 12.3816. *(4 places)*

_____ **36.** A contractor tiles a 15.5′ × 13.8′ family room and a 22.3′ × 5.0′ hallway. What is the total area tiled?

_____ **37.** A laborer works 7.86 hr, 5.83 hr, and 6.75 hr installing windows. At $10.50 per hour, how much money did the laborer gross?

_____ **38.** 6.08 × 4.29 *(4 places)*

_____ **39.** 83.6 ÷ $\frac{2}{5}$

_____ **40.** Change six thousand, two hundred forty-six dollars and thirty seven cents to a decimal number.

_____ **41.** What is the product of 7.6 × 2.81 × 4.5? *(3 places)*

_____ **42.** A 121.5′ cable is divided into nine equal pieces. How long is each segment?

_____ **43.** 79.62 ÷ .4

Name _____ Date _____

1. 34 ÷ .8

2. 34.9805 × 1.156 *(6 places)*

3. 37.03 + .521 + .9 + 1000 + 4000.0014 *(4 places)*

4. 34.895 × .56 *(4 places)*

5. 900 − .009 *(3 places)*

6. Change ²⁄₂₅ to a decimal.

7. Change .750 to a fraction.

8. Change .90 to fractional 10ths.

9. Subtract twenty-two ten thousandths from 10.0302. *(3 places)*

10. What is the sum of twenty-six and twenty-six hundredths + seven tenths + six and eighty-three thousandths + four and seven thousandths?

11. In estimating an interior painting job, ceilings and walls of five rooms have net areas of (after subtracting all openings) 190.8 sq yd, 162.6 sq yd, 128.5 sq yd, 202.4 sq yd, and 98.2 sq yd. What is the total area to be painted?

12. What is the cost of three bolts of cloth each containing 36.5 yd at $6.50 per yard?

13. An electrical contractor took a wiring job for $12,000.00. Five electricians and five helpers are hired to do the job. They work 8 hours a day. The electricians are paid $15.25 per hour, and the helpers are paid $7.25 per hour. The contractor spent $4200.00 on materials. In how many workdays must the job be completed so that the contractor clears $2400.00?

14. Manganese bronze contains the following amounts of metals per pound: copper .89 lb, tin .10 lb, and manganese .01 lb. How much copper is in a shaft that weighs 235 lb?

15. The full-load current of a 1 HP, 230 V, 3ɸ, AC induction motor is 3.6 A. What is the full-load current as a fraction?

16. A type S, size 4 steel bar has a diameter of .250″. What is the diameter as a fraction?

PERCENTAGES 6

A percentage expresses a part of a whole number in terms of hundredths. A whole quantity can be divided into any number of equal parts, such as fractions or decimals. Simple interest, discounts, commissions, and taxes are percentages used in business transactions.

PERCENTAGE

A *percentage* is a method of expressing a part of a whole number in terms of hundredths. A *percent* is one part of 100 parts. The percent sign (%) indicates the number is a part of 100. See Figure 6-1. For example, a copper casting alloy (100%) consists of 88% copper (88 parts of 100), 10% tin (10 parts of 100), and 2% zinc (2 parts of 100).

Fraction and Decimal Equivalents

A whole quantity can be divided into any number of equal parts. The number 1 represents one whole unit. All fractions are part of one whole unit. Fractions with the same numerator and denominator are equal to 1. For example, in fractions, 1 can be $\frac{2}{2}$, $\frac{10}{10}$, $\frac{25}{25}$, $\frac{100}{100}$, etc. In percentages, one whole unit is 100 equal parts, $\frac{100}{100}$, or 100%. Any percentage, such as 7%, 10%, 25%, 55%, etc., which is less than 100%, is less than one whole unit. Written as decimals, these percentages have a decimal point in front of the number, such as .07, .10, .25, .55, etc.

All fractions and decimals have equivalent percentages. See Appendix. For example, the fraction $\frac{1}{2}$ has a decimal equivalent of .5, and .5 has an equivalent percentage of 50% ($\frac{1}{2}$ = **.5 or 50%**).

A decimal number having two places to the right of the decimal point occupies the hundredths place. To change a decimal to a percentage, multiply by 100 (move decimal point two places to the right). For example, to change .25 to a percentage, multiply .25 by 100 (.25 × 100 = **25%**).

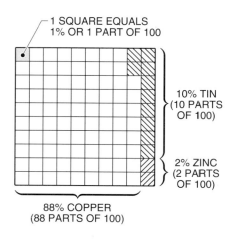

Figure 6-1. A percentage is a method of expressing a part of a whole number in terms of hundredths.

To change a percentage to a decimal number, divide by 100 (move the decimal point two places to the left). For example, to change 75% to a decimal number, divide 75 by 100 (75 ÷ 100 = .75).

A *fractional hundredth* is a fraction having a denominator of 100 and a numerator equal to the percent. For example, to change 70% to fractional hundredths, place 70 (number of hundredths) over 100 (70% = $^{70}/_{100}$).

EXAMPLES

Fraction and Decimal Equivalents

1. Change .54 to a percentage.

❶ $\left\{ \right.$ **.54 x 100 = 54%**

❶ MULTIPLY DECIMAL BY 100

① Multiply .54 by 100 (.54 × 100 = **54%**).

2. Change 71% to a decimal.

❶ $\left\{ \dfrac{71\%}{100} = .71 \right.$

❶ DIVIDE PERCENTAGE BY 100

① Divide 71 by 100 (71 ÷ 100 = **.71**).

3. Change 35% to fractional hundredths.

35% = $\dfrac{35}{100}$ $\left. \right\}$ ❶

❶ PLACE PERCENTAGE OVER 100

① Place 35 over 100 (35% = $^{35}/_{100}$).

Mixed Number Percentages. To change percentages that are mixed numbers to proper fractions, change the mixed number to an improper fraction, and place over 100. See Figure 6-2. Reduce by dividing the improper fraction (numerator) by 100 (denominator), and cancelling as required.

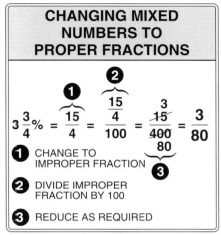

Figure 6-2. To change mixed number percentages to proper fractions, change the percentage to an improper fraction and divide by 100.

EXAMPLES

Mixed Numbers

1. Change $43\frac{3}{4}\%$ to a proper fraction.

$$43\frac{3}{4}\% \overset{\text{①}}{=} \frac{175}{4} \overset{\text{②}}{=} \frac{\frac{175}{4}}{100} = \frac{\cancel{175}^{35}}{\cancel{400}_{80}} = \frac{7}{16}$$

① CHANGE TO IMPROPER FRACTION
② DIVIDE IMPROPER FRACTION BY 100
③ REDUCE AS REQUIRED

① Change $43\frac{3}{4}$ to $^{175}/_4$. ② Place over 100 ($^{175/4}/_{100}$). ③ Reduce and cancel $^{175/4}/_{100}$ by dividing $^{175}/_4$ by 100 ($^{175}/_4 \div 100 = ^{175}/_4 \times ^{1}/_{100} = ^{175}/_{400} = $ **$^{7}/_{16}$**).

2. Change $38\frac{3}{4}\%$ to a proper fraction.

$$38\frac{3}{4}\% \overset{\text{①}}{=} \frac{155}{4} \overset{\text{②}}{=} \frac{\frac{155}{4}}{100} = \frac{\cancel{155}^{31}}{\cancel{400}_{80}} = \frac{31}{80}$$

① CHANGE TO IMPROPER FRACTION
② DIVIDE IMPROPER FRACTION BY 100
③ REDUCE AS REQUIRED

① Change $38\frac{3}{4}$ to $^{155}/_4$. ② Place over 100 ($^{155/4}/_{100}$). ③ Reduce and cancel $^{155/4}/_{100}$ by dividing $^{155}/_4$ by 100 ($^{155}/_4 \div 100 = ^{155}/_4 \times ^{1}/_{100} = 155/400 = $ **$^{31}/_{80}$**).

Percentages Larger Than 100. Any percentage larger than 100% is more than one whole unit. For example, 110% and 125% are more than one whole unit. Percentages smaller than 100% are less than one whole unit. Percentages larger than 100% are one whole unit and a part of another added together. See Figure 6-3.

To change 110% to a decimal, divide by 100 (110 ÷ 100 = 1.10). To change the decimal number to a fraction, multiply .10 (decimal number) by 100 (move decimal point two places to the left), and place over 100 (.10 × 100 = 10; $^{10}/_{100} = ^{1}/_{10}$). Add the fraction to 1 (110% = **1.10 = $1^{1}/_{10}$**).

CHANGING PERCENTAGES LARGER THAN 100% TO DECIMALS AND FRACTIONS

①   **②**

$$120\% = \frac{\overbrace{120}}{100} = 1.20 = 1\frac{\overbrace{20}}{100}$$

③ $\left\{ = 1\frac{\overbrace{\frac{1}{20}}}{\underset{5}{100}} = 1\frac{1}{5} \right.$

① DIVIDE BY 100

② DIVIDE DECIMAL NUMBER (AS WHOLE NUMBER) BY 100

③ REDUCE AS REQUIRED

Figure 6-3. To change percentages larger than 100% to decimals and fractions, divide the percentage by 100, divide the decimal by 100, and reduce.

EXAMPLES

Percentages Larger Than 100

1. Change 125% to a decimal and a fraction.

①   **②**

$$125\% = \frac{\overbrace{125}}{100} = 1.25 = 1\frac{\overbrace{25}}{100}$$

③

$$= 1\frac{\overbrace{\frac{1}{25}}}{\underset{4}{100}} = 1\frac{1}{4}$$

① DIVIDE BY 100

② PLACE DECIMAL OVER 100

③ REDUCE AS REQUIRED

① Divide 125 by 100 (125 ÷ 100 = 1.25). ② Multiply .25 by 100 and place over 100 (.25 × 100 = 25; $^{25}/_{100}$ = ¼). ③ Reduce as required (125% = **1.25** = **1¼**).

2. Change 330% to a decimal and a fraction.

①   **②**

$$330\% = \frac{\overbrace{330}}{100} = 3.30 = 3\frac{\overbrace{30}}{100}$$

③

$$= 3\frac{\overbrace{\frac{3}{30}}}{\underset{10}{100}} = 3\frac{3}{10}$$

① DIVIDE BY 100

② PLACE DECIMAL OVER 100

③ REDUCE AS REQUIRED

① Divide 330 by 100 (330 ÷ 100 = 3.30). ② Multiply .30 by 100 and place over 100 (.30 × 100 = 30; $^{30}/_{100}$ = $^{3}/_{10}$). ③ Reduce as required (330% = **3.30** = **3³⁄₁₀**).

Rounding Decimals. A decimal number that has more than two places to the right of the decimal point can occur when calculating percentages. In this case, the decimal number is rounded to the second decimal place, or hundredths. See Figure 6-4.

To round a decimal number to the hundredths (second decimal place), add 1 to the hundredths (round up) if the number in the thousandths (third decimal place) is

5, 6, 7, 8, or 9, and drop the rest of the decimal number.

If the number in the thousandths is 0, 1, 2, 3, or 4, leave the hundredths the same (round down). Drop the rest of the decimal number.

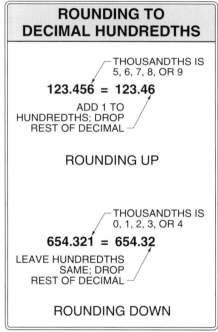

ROUNDING TO DECIMAL HUNDREDTHS

THOUSANDTHS IS 5, 6, 7, 8, OR 9

123.456 = 123.46

ADD 1 TO HUNDREDTHS; DROP REST OF DECIMAL

ROUNDING UP

THOUSANDTHS IS 0, 1, 2, 3, OR 4

654.321 = 654.32

LEAVE HUNDREDTHS SAME; DROP REST OF DECIMAL

ROUNDING DOWN

Figure 6-4. Decimal numbers can be rounded to any decimal place depending on the degree of accuracy required.

Decimal numbers can be rounded to any decimal place depending on the required degree of accuracy. For example, to round 35.7168 to the thousandths, add 1 to the 6 (thousandths), because 8 occupies the ten thousandths. Drop the rest of the decimal. The number 35.7168 rounded to the thousandths is **35.717**.

PRACTICE PROBLEMS

Fraction and Decimal Equivalents

1. Change $\frac{1}{4}$ to a percentage.
2. Change 325% to a fraction.
3. Change 14% to a fractional hundredth.
4. Change .22 to a percentage.
5. Change 275% to a decimal.
6. Change A, B, C, D, and E on the Pie Chart to decimals.

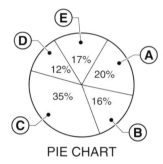

PIE CHART

Round the following decimals to the hundredths.

7. 134.9845
8. .9094
9. .5983
10. 1.005

Percentages of Quantities

To find a percentage of a quantity, change the percentage to a decimal number, and multiply the quantity by the decimal number. See Figure 6-5. For example, to find 25% of 400, change 25% to a decimal (25 ÷ 100 = .25). Multiply 400 by .25 (400 × .25 = **100**).

FINDING PERCENTAGE OF QUANTITIES

❶

30% OF 240 = 240 x .30 = 72

❶ MULTIPLY BY PERCENTAGE
(DECIMAL FORM)

Figure 6-5. To find the percentage of a quantity, multiply the quantity by the percentage in decimal form.

EXAMPLES

Percentages of Quantities

1. Find 8% of $245.50.

❶

8% OF $245.50 = 245.50 x .08
= $19.64

❶ MULTIPLY BY PERCENTAGE
(DECIMAL FORM)

① Change 8% to a decimal (8 ÷ 100 = .08). Multiply 245.50 by .08 (245.50 × .08 = **$19.64**).

2. Find 2.7% of 54.

❶

2.7% OF 54 = 54 x .027 = 1.458

❶ MULTIPLY BY PERCENTAGE
(DECIMAL FORM)

① Change 2.7% to a decimal (2.7 ÷ 100 = .027). Multiply 54 by .027 (54 × .027 = **1.458**).

3. Find 135% of 750.

❶

135% OF 750 = 750 x 1.35
= 1012.5

❶ MULTIPLY BY PERCENTAGE
(DECIMAL FORM)

① Change 135% to a decimal (135 ÷ 100 = 1.35). Multiply 750 by 1.35 (750 × 1.35 = **1012.5**).

4. Find $21\frac{7}{8}\%$ of 240.

❶

$21\frac{7}{8}$% OF 240 = 240 x .21875
= 52.5

❶ MULTIPLY BY PERCENTAGE
(DECIMAL FORM)

① Change $21\frac{7}{8}\%$ ($\frac{7}{8} = 7 ÷ 8 = .875$) to a decimal (21.875 ÷ 100 = .21875). Multiply 240 by .21875 (240 × .21875 = **52.5**).

PRACTICE PROBLEMS

Percentages of Quantities

1. Find 20% of 35.
2. Find 22% of 40.
3. Find $66\frac{2}{3}\%$ of 500.
4. Find 12% of 2125.
5. Find 125% of 355.
6. Find 8.7% of 135.
7. Find 130% of 100.
8. Find 25% of 330.
9. Find 110% of 200.
10. Find $33\frac{1}{3}\%$ of 1000.
11. A salesperson receives 12% of total sales. If total sales are $1965.00, how much does the salesperson receive?

12. A book publisher printed 8000 copies of a basic text on mathematics. The spoilage at the printer and the bindery reached 9% of the total run. How many of the books were spoiled?

Finding Percentages

To find what percentage one number is of another, divide the percent number by the other number. Multiply the quotient by 100 (move decimal two places to the right) to find the percentage. See Figure 6-6.

For example, to find what percentage 8 is of 40, divide 8 (percent number) by 40 ($8 \div 40 = .2$). Multiply .2 by 100 ($.2 \times 100 = \mathbf{20\%}$).

To simplify the problem, before dividing the percent number by the other number, reduce as required (as in reducing fractions). For example, to find what percentage 5 is of 20, divide 5 by 20 or $\frac{5}{20}$. Reduce $\frac{5}{20}$ to lowest terms by dividing 5 and 20 by 5 ($\frac{5 \div 5}{20 \div 5} = \frac{1}{4}$). Find $\frac{1}{4}$ on the Equivalents Table, or divide 1 by 4 ($1 \div 4 = .25$). Multiply .25 by 100 ($.25 \times 100 = \mathbf{25\%}$).

PERCENTAGE ONE NUMBER IS OF ANOTHER

PERCENT NUMBER ❷

3 OF 12 = 3 ÷ 12 = .25 x 100
❶
= 25%

❶ DIVIDE PERCENT NUMBER BY OTHER NUMBER

❷ MULTIPLY BY 100

OR

3 OF 12 = $\frac{3}{12}$ = $\frac{1}{4}$ = 25%
❶ ❷

❶ REDUCE TO LOWEST TERMS

❷ FIND PERCENT EQUIVALENT

Figure 6-6. To find the percentage one number is of another, divide and multiply by 100 or reduce to lowest terms.

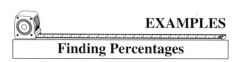

EXAMPLES

Finding Percentages

1. What percentage of 50 is 4?

❶
____ % OF 50 IS 4 = $\frac{4}{50}$

= .08 x 100 = 8%
❷

❶ DIVIDE PERCENT NUMBER BY OTHER NUMBER

❷ MULTIPLY BY 100

① Divide 4 (percent number) by 50 ($4 \div 50 = .08$). ② Multiply .08 by 100 ($.08 \times 100 = 08 = \mathbf{8\%}$).

2. What percentage of 120 is 90?

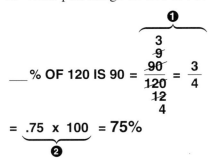

① DIVIDE PERCENT NUMBER BY OTHER NUMBER AND REDUCE AS REQUIRED

② MULTIPLY BY 100

① Divide 90 by 120 or $^{90}/_{120}$. Reduce $^{90}/_{120}$ to lowest terms by dividing 90 and 120 by 10 ($\frac{90 \div 10}{120 \div 10} = \frac{9}{12}$). Reduce $^{9}/_{12}$ by dividing 9 and 12 by 3 ($\frac{9 \div 3}{12 \div 3} = \frac{3}{4}$). Find $^{3}/_{4}$ on the Equivalents Table, or divide 3 by 4 ($3 \div 4 = .75$). ② Multiply .75 by 100 ($.75 \times 100 = $ **75%**).

PRACTICE PROBLEMS

Finding Percentages

1. What percentage of 130 is 19.50?

2. What percentage of 500 is 200?

3. What percentage of 500 is 250?

4. A statute mile is 5280′ and a nautical mile is 6076′. What percentage of a nautical mile is a statute mile?

5. A college student who worked during the summer earned $960.00 and saved $640.00. What percentage of the summer earnings is saved?

6. Seventy bolts are defective in a box of 840 bolts. What percentage of the bolts is defective?

7. A school had 720 students. Enrollment increased by 108 students in six months. What is the percentage of increase in enrollment?

8. A contractor agreed to build a 20 mile road. What percentage of the work is completed after 15 miles are finished?

9. A building is to be 40 stories high. What percent of the height is completed after 16 stories are finished?

10. A community of 14,000 has a public high school enrollment of 1800. What percent of the community's population goes to the public high school?

Numbers Using Percentages

To find a number when a number and percentage are given, divide the given number by the given percentage (decimal form). See Figure 6-7. For example, to find what number 7 is 25% of, divide 7 by .25 ($7 \div .25 = $ **28**).

FINDING A NUMBER GIVEN A PERCENTAGE

12 IS 75% OF ___ = $\overbrace{12 \div .75}^{1}$

$$= 16$$

1 DIVIDE NUMBER BY PERCENTAGE (DECIMAL FORM)

Figure 6-7. To find a number when a number and a percentage are given, divide the given number by the percentage in decimal form.

EXAMPLES
Numbers Using Percentages

1. Eight is 20% of what number?

8 IS 20% OF ___ = $\dfrac{\overbrace{8}^{1}}{.20}$ = 40

1 DIVIDE NUMBER BY PERCENTAGE (DECIMAL FORM)

① Divide 8 by .20 (8 ÷ .20 = **40**).

2. Two is 50% of what number?

2 IS 50% OF ___ = $\dfrac{\overbrace{2}^{1}}{.50}$ = 4

1 DIVIDE NUMBER BY PERCENTAGE (DECIMAL FORM)

① Divide 2 by .50 (2 ÷ .50 = **4**).

PRACTICE PROBLEMS
Numbers Using Percentages

1. One is 10% of what number?
2. Sixteen is 4% of what number?

3. Twenty is 5% of what number?
4. Thirty-eight is 2% of what number?
5. Twenty-seven is 9% of what number?
6. One hundred twenty is 15% of what number?
7. What is the horsepower (HP) of a 78% efficient motor that delivers 23.4 HP?
8. A benefit campaign collects $5000.00 in one week, which is 50% of the total contributions. What are the total contributions?
9. A transformer has a 320 watt (W) rating. What percent of the rated power is lost by a 48 W loss?
10. An ironworker fabricates 120′ of railing, which is 30% of an order. How much railing is needed to complete the order?

BUSINESS TRANSACTIONS

Percentages are used in many business transactions. Percentages are used to find simple interest, discounts, commissions, and taxes.

Simple Interest

Simple interest is the amount of money charged for money borrowed (loan). Simple interest is also known as interest. *Principal* is the amount of money borrowed. *Interest rate* is the percent of the principal paid back. The *time* is the amount of time allotted for repayment of the principal plus interest. Simple interest is dependent on the principal, interest rate, and time.

Finding Simple Interest. To find simple interest, multiply the principal by the interest rate (decimal form), and multiply by the time. See Figure 6-8.

To find the interest for a loan of $2000.00 for one year, at an interest rate of 13%, multiply 2000 (principal) by .13 (2000 × .13 = 260). Multiply 260 by one year (260 × 1 = **$260.00**). If the time of this loan is two years, then the interest is **$520.00** (260 × 2 = 520).

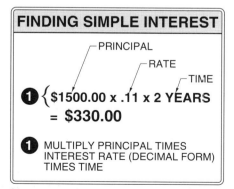

FINDING SIMPLE INTEREST

PRINCIPAL

RATE

TIME

❶ { **$1500.00 x .11 x 2 YEARS = $330.00**

❶ MULTIPLY PRINCIPAL TIMES
INTEREST RATE (DECIMAL FORM)
TIMES TIME

Figure 6-8. To find simple interest, multiply the principal times the interest rate and multiply by the time.

| Finding Simple Interest |

1. Find the interest at 5% on $850.00 for one year.

❶

$850.00 x .05 x 1 YEAR = $42.50

❶ MULTIPLY PRINCIPAL TIMES
INTEREST RATE (DECIMAL FORM)
TIMES TIME

① Multiply 850 by .05 (850 × .05 = 42.5), and multiply 42.5 by one year (42.5 × 1 = **$42.50**).

2. Find the interest at 6% on $1250.00 for five years.

❶

$1250.00 x .06 x 5 YEARS = $375.00

❶ MULTIPLY PRINCIPAL TIMES
INTEREST RATE (DECIMAL FORM)
TIMES TIME

① Multiply 1250 by .06 (1250 × .06 = 75), and multiply 75 by five years (75 × 5 = **$375.00**).

Finding Principal. To find the principal when interest, interest rate, and time are known, find the interest for one year (divide total interest by number of years), and divide by the interest rate (decimal form). See Figure 6-9. For example, to find the principal for a loan at 11% for four years and interest of $1540.00, divide 1540 by 4 (1540 ÷ 4 = 385).

Divide 385 by .11 (385 ÷ .11 = 3500 = **$3500.00**).

FINDING PRINCIPAL

❶ { **$1200.00 ÷ 5 YEARS**

❷ { **= 240.00 ÷ .10**

= $2400.00 ⟍INTEREST RATE

❶ DIVIDE TOTAL
INTEREST BY TIME

❷ DIVIDE INTEREST BY INTEREST
RATE (DECIMAL FORM)

Figure 6-9. To find principal, divide the total interest by the time, and divide by the interest rate.

EXAMPLE
Finding Principal

1. Find the principal if the interest for three years at 5% is $52.50.

$$\overset{\textbf{❶}}{\frac{\$52.50}{3\ \text{YEARS}}} = \underset{\textbf{❷}}{17.50 \div .05}$$

= $350.00

❶ DIVIDE TOTAL INTEREST BY TIME
❷ DIVIDE INTEREST BY
INTEREST RATE (DECIMAL FORM)

① Divide 52.50 by 3 (52.50 ÷ 3 = 17.50). ② Divide 17.50 by .05 (17.50 ÷ .05 = **$350.00**).

Finding Interest Rate. To find the interest rate when principal, time, and interest are known, find interest for one year (divide total interest by number of years), divide by the principal, and multiply by 100. See Figure 6-10. For example, to find the interest rate for a $1500.00 loan for two years, with interest of $180.00, divide 180 by 2 (180 ÷ 2 = 90). Divide 90 by 1500 (90 ÷ 1500 = .06). Multiply .06 by 100 (.06 × 100 = **6%**).

FINDING INTEREST RATE

❶ { **$300.00 ÷ 3 YEARS**

❷ { **= 100.00 ÷ 2500.00**

❸ { **= .04 x 100 = 4%**

❶ DIVIDE TOTAL
INTEREST BY TIME

❷ DIVIDE QUOTIENT
BY PRINCIPAL

❸ MULTIPLY QUOTIENT BY 100

Figure 6-10. To find interest rate, divide the total interest by the time, divide by the principal, and multiply by 100.

EXAMPLE
Finding Interest Rate

1. Find the interest rate if the interest on a $1120.00 loan for three years and six months is $156.80.

❶ $\{$ **$156.80 ÷ 3.5 YEARS**

❷ $\{$ **= 44.8 ÷ 1120**

❸ $\{$ **= .04 x 100 = 4%**

❶ DIVIDE TOTAL INTEREST BY TIME
❷ DIVIDE BY PRINCIPAL
❸ MULTIPLY BY 100

① Divide 156.80 by 3.5, three years and six months (156.80 ÷ 3.5 = 44.8). ② Divide 44.8 by 1120 (44.8 ÷ 1120 = .04). ③ Multiply .04 by 100 (.04 × 100 = 04 = **4%**).

Finding Time. To find the time when principal, interest, and rate are known, multiply the principal by the interest rate (decimal form). Divide total interest by the quotient. See Figure 6-11.

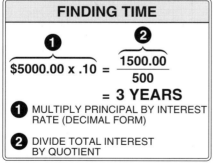

FINDING TIME

❶ ❷

$5000.00 x .10 = $\dfrac{1500.00}{500}$

= 3 YEARS

❶ MULTIPLY PRINCIPAL BY INTEREST RATE (DECIMAL FORM)
❷ DIVIDE TOTAL INTEREST BY QUOTIENT

Figure 6-11. To find time, multiply the principal by the interest rate, and divide the total interest by the quotient.

For example, to find the time for $200.00 to gain $45.00 at a rate of 5%, multiply 200 by 5 (200 × 5 =

1000). Divide 1000 by 100 (1000 ÷ 100 = 10). Divide 45 by 10 (45 ÷ 10 = 4.5 = **4.5 years** or **four years and six months**).

EXAMPLE

Finding Time

1. Find the time for $275.00 to gain $55.00 at a rate of 6%.

❶ $\{$ **$275.00 x .06 = 16.5**

❷ $\{$ **55 ÷ 16.5 = 3.$\overline{33}$**

= 3 YEARS 4 MONTHS

❶ MULTIPLY PRINCIPAL BY RATE (DECIMAL FORM)
❷ DIVIDE TOTAL INTEREST BY PRODUCT

① Multiply 275 by .06 (275 × .06 = 16.5). ② Divide 55 by 16.5 (55 ÷ 16.5 = 3.$\overline{33}$ = **3.$\overline{33}$ years** or **three years and four months**).

PRACTICE PROBLEMS

Simple Interest

1. Find the interest on $500.00 at 6% for one year.
2. Find the interest on $800.00 at 4% for one year.
3. Find the interest on $1500.00 at 5% for two years.
4. Find the interest on $347.00 at 3% for three years.
5. Find the interest on $470.00 at 6% for six months.

6. Find the interest on $1200.00 at 3% for eight months.

7. Find the interest on $900.00 at $4\frac{1}{2}$% for three months.

8. At what rate of interest does the principal earn $225.00?

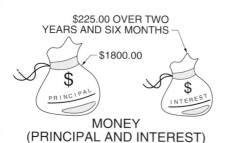

$225.00 OVER TWO YEARS AND SIX MONTHS

$1800.00

$ PRINCIPAL

$ INTEREST

MONEY
(PRINCIPAL AND INTEREST)

9. At what rate of interest does $5000.00 gain $1100.00 interest in five years and six months?

10. At what rate of interest does $750.00 gain $125.00 interest in eight years and four months?

11. At what rate of interest does $4050.00 gain $1336.50 interest in six years?

12. What principal gains $15.00 interest in six months at 3%?

13. What principal gains $96.00 interest in three years at 4%?

14. What principal gains $195.75 interest in three years at $4\frac{1}{2}$%?

15. In what time does $6000.00 gain $1680.00 interest at 4%?

16. In what time does $1800.00 gain $504.00 interest at 3%?

17. In what time does $25,000.00 gain $4125.00 interest at $5\frac{1}{2}$%?

18. At what rate of interest does $2150.00 gain $150.50 interest in two years?

Discount

Marked price is the price at which products are originally intended to be sold. *Discount* is the reduction (amount of money) of the marked price. *Net price* is the amount remaining after the discount is subtracted from the marked price.

Finding Discount. To find the discount, multiply the marked price by the percent discount (decimal form). See Figure 6-12. For example, to find the discount of a product with a marked price of $25.00, at a percent discount of 20%, multiply 25 by .20 (25 × .20 = **$5.00**).

FINDING DISCOUNT

1 $50.00 x .20 = **$10.00**

.20 = 20%

1 MULTIPLY MARKED PRICE BY PERCENT DISCOUNT (DECIMAL FORM)

Figure 6-12. To find discount, multiply the marked price by the percent discount.

When a discount has more than two decimal places, round the decimal to the nearest cent (hundredths). For example, to find the discount of a product with a marked price of $10.25, at a percent discount of 15%, multiply 10.25 by .15 (10.25 × .15 = 1.5375 = 1.54 rounded = **$1.54**).

EXAMPLE

Finding Discount

1. Find the discount on a product with a marked price of $122.00, at a percent discount of 60%.

❶ { $122.00 x .60 = **$73.20**

❶ MULTIPLY MARKED PRICE
BY PERCENT DISCOUNT
(DECIMAL FORM)

① Multiply 122 by .60 (122 × .60 = **$73.20**).

Finding Net Price. To find the net price when the discount and the marked price are known, subtract the discount from the marked price. See Figure 6-13.

FINDING NET PRICE

❶

$100.00 − $30.00 = **$70.00**

❶ SUBTRACT DISCOUNT
FROM MARKED PRICE

Figure 6-13. To find net price, subtract the discount from the marked price.

For example, to find the net price of a product with a marked price of $25.00, and a discount of

$6.50, subtract 6.50 from 25.00 (25.00 − 6.50 = **$18.50**).

EXAMPLE

Finding Net Price

1. Find the net price of a product with a marked price of $144.00 and a discount of $24.50.

❶ { $144.00 − $24.50 = **$119.50**

❶ SUBTRACT DISCOUNT
FROM MARKED PRICE

① Subtract 24.50 from 144.00 (144.00 − 24.50 = 119.50 = **$119.50**).

Finding Percent Discount. To find the percent discount if the discount and the marked price are known, divide the discount by the marked price, and multiply the quotient by 100 (move decimal two places to the right). See Figure 6-14. For example, to find the percent discount of a product with a marked price of $25.00, and a discount of $6.50, divide 6.50 by 25.00 (6.50 ÷ 25.00 = .26). Multiply .26 by 100 (.26 × 100 = **26%**).

To find the percent discount if the discount price and the marked price are known, divide the discount price by the marked price, multiply the quotient by 100 (move decimal two places to the right), and subtract from 100. For example, to find the percent discount of a product with a marked price of $25.00 and a discount price of $21.25, divide 21.25 by 25.00 (21.25 ÷ 25.00 = .85). Multiply .85 by 100 (.85 × 100 = 85). Subtract 85 from 100 (100 − 85 = **15%**).

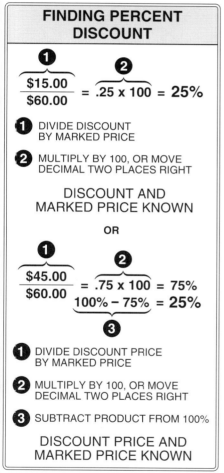

FINDING PERCENT DISCOUNT

❶   ❷

$$\frac{\$15.00}{\$60.00} = \overbrace{.25 \times 100} = \mathbf{25\%}$$

❶ DIVIDE DISCOUNT BY MARKED PRICE

❷ MULTIPLY BY 100, OR MOVE DECIMAL TWO PLACES RIGHT

DISCOUNT AND MARKED PRICE KNOWN

OR

❶   ❷

$$\frac{\$45.00}{\$60.00} = \overbrace{.75 \times 100} = 75\%$$
$$\underbrace{100\% - 75\%} = \mathbf{25\%}$$
❸

❶ DIVIDE DISCOUNT PRICE BY MARKED PRICE

❷ MULTIPLY BY 100, OR MOVE DECIMAL TWO PLACES RIGHT

❸ SUBTRACT PRODUCT FROM 100%

DISCOUNT PRICE AND MARKED PRICE KNOWN

Figure 6-14. Percent discount can be found if the discount and the marked price are known, or if the discount price and the marked price are known.

EXAMPLES

Finding Percent Discount

1. Find the percent discount of a product with a marked price of $40.00 and a discount of $10.00.

❶   ❷

$$\overbrace{\$10.00 \div \$40.00} = \overbrace{.25 \times 100}$$
$$= \mathbf{25\%}$$

❶ DIVIDE DISCOUNT BY MARKED PRICE

❷ MULTIPLY QUOTIENT BY 100

① Divide 10 by 40 (10 ÷ 40 = .25). ② Multiply .25 by 100 (.25 × 100 = **25%**).

2. Find the percent discount of a product with a marked price of $220.00 and a discount price of $165.00.

❶   ❷

$$\overbrace{\$165.00 \div \$220.00} = \overbrace{.75 \times 100}$$
$$= \mathbf{75\%}$$
$$\underbrace{100\% - 75\%} = \mathbf{25\%}$$
❸

❶ DIVIDE DISCOUNT PRICE BY MARKED PRICE

❷ MULTIPLY QUOTIENT BY 100

❸ SUBTRACT PRODUCT FROM 100%

① Divide 165 by 220 (165 ÷ 220 = .75). ② Multiply .75 by 100 (.75 × 100 = 75). ③ Subtract 75 from 100 (100 − **75 = 25%**).

Finding Marked Price. To find the marked price, divide discount by percent discount (decimal form). See Figure 6-15. For example, to find the marked price of a product with a discount of $35.00, and a percent discount of 12%, divide 35.00 by .12 (35.00 ÷ .12 = 291.666 = **$291.67**).

FINDING MARKED PRICE

❶ $\{$ **$20.00 ÷ .25 = $80.00**

❶ DIVIDE DISCOUNT BY PERCENT DISCOUNT (DECIMAL FORM)

Figure 6-15. To find marked price, divide the discount by the percent discount.

EXAMPLE

Finding Marked Price

1. Find the marked price for a product with a discount of $16.00 and a percent discount of 40%.

❶ $\{$ **$16.00 ÷ .40 = $40.00**

❶ DIVIDE DISCOUNT BY PERCENT DISCOUNT (DECIMAL FORM)

① Divide 16 by .40 (16 ÷ .40 = **$40.00**).

Finding Multiple Discounts. A *multiple discount* is the discount found by applying more than one percent discount to the marked price, resulting in the net price. To find the net price of a product with a multiple discount, apply the first percent discount to the marked price to find the first discount price. Apply the second percent discount

to the first discount price to find the second discount price, etc. See Figure 6-16.

Figure 6-16. To find multiple discount, apply the percent discount to the marked price. To find the second discount price, apply the second percent discount to the first discount price, etc.

For example, to find the net price of a product with a marked price of $75.40, a percent discount of 20%, and an additional percent discount of 10%, multiply 75.40 by .20 (75.40 × .20 = 15.08). Subtract

15.08 from 75.40 (75.40 − 15.08 = **$60.32**). Multiply 60.32 by .10 to find the second discount (60.32 × .10 = 6.032 = 6.03, rounded). Subtract 6.03 from 60.32 (60.32 − 6.03 = **$54.29**).

③ Multiply 487.50 by .10 to find the second discount (487.50 × .10 = 48.75). ④ Subtract 48.75 from 487.50 (487.50 − 48.75 = **$438.75**).

EXAMPLE

Finding Multiple Discounts

1. Find the net price of a product with a marked price of $650.00, and a multiple discount of 25% and 10%.

❶   ┌ FIRST
       │ DISCOUNT
$650.00 x .25 = 162.50

❷   ┌ FIRST
       │ DISCOUNT
       │ PRICE
650.00 − 162.50 = 487.50

❸   ┌ SECOND
       │ DISCOUNT
487.50 x .10 = 48.75

❹   ┌ NET PRICE
487.50 − 48.75 = $438.75

❶ MULTIPLY MARKED PRICE BY FIRST PERCENT DISCOUNT (DECIMAL FORM)

❷ SUBTRACT FIRST DISCOUNT FROM MARKED PRICE

❸ MULTIPLY FIRST DISCOUNT PRICE BY SECOND PERCENT DISCOUNT (DECIMAL FORM)

❹ SUBTRACT SECOND DISCOUNT FROM FIRST DISCOUNT

① Multiply 650.00 by .25 to find the first discount (650.00 × .25 = 162.50). ② Subtract 162.50 from 650.00 to find the first discount price (650.00 − 162.50 = **$487.50**).

PRACTICE PROBLEMS

Discount

1. Find the discount of a product with a marked price of $22,000.00 at a percent discount of 5%.

2. What is the percent discount of the Bricks?

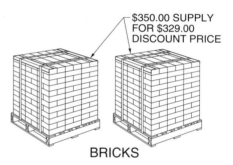

$350.00 SUPPLY FOR $329.00 DISCOUNT PRICE

BRICKS

3. Find the net price of a product with a marked price of $250.00 and a multiple discount of 10% and 5%.

4. Merchandise is purchased for $1250.00. The merchant is offered a 3½% discount for cash. What is the cash discount?

5. Which is more profitable for a buyer on a $3500.00 purchase, a single discount of 15%, or a multiple discount of 10% and 5%?

6. A purchase of $8920.00 has a discount of $312.20. What is the percent discount?

7. A building contractor buys $3550.00 of building supplies from a building supply discount warehouse for a discount price of $2946.50. What is the percent discount?

Commission

A *commission* is a percentage of the total sale (total money received for a business transaction), which is paid to the individual who performs the transaction. *Commission rate* is the percentage used to figure commissions. For example, if a salesperson receives 3% of the selling price of a car, then 3% is the commission rate.

Net proceeds is the amount of the total sale left after the commission is deducted. For example, if $100.00 is the total sale, and the commission is $12.00, then the net proceeds is $88.00 (100 − 12 = **88**).

Finding Commission. To find commission, multiply the total sale by the commission rate (decimal form). See Figure 6-17. For example, to find the commission made on a $13,600.00 sale with a 10% commission rate, multiply 13,600 by .10 (13,600 × .10 = **$1360.00**).

FINDING COMMISSION

❶

$11,000.00 x .08 = **$880.00**

❶ MULTIPLY TOTAL SALE BY COMMISSION RATE (DECIMAL FORM)

Figure 6-17. To find commission, multiply the total sale by the commission rate.

EXAMPLE

Finding Commission

1. Find the commission made on a $450.00 sale with a 12% commission rate.

❶ {$450.00 x .12 = **$54.00**

❶ MULTIPLY TOTAL SALE BY COMMISSION RATE (DECIMAL FORM)

① Multiply 450 by .12 (450 × .12 = **$54.00**).

Finding Net Proceeds. To find net proceeds, subtract commission from the total sale. See Figure 6-18.

FINDING NET PROCEEDS

❶

$150.00 − $18.00 = **$132.00**

❶ SUBTRACT COMMISSION FROM TOTAL SALE

Figure 6-18. To find net proceeds, subtract commission from the total sale.

For example, to find net proceeds for a $13,600.00 sale for

which a $1360.00 commission is paid, subtract 1360 from 13,600 (13,600 − 1360 = **$12,240.00**).

EXAMPLE
Finding Net Proceeds

1. Find the net proceeds for a $250.00 sale for which a $60.00 commission is paid.

❶ { **$250.00 − $60.00 = $190.00**

❶ SUBTRACT COMMISSION FROM TOTAL SALE

① Subtract 60 from 250 (250 − 60 = **$190.00**).

Finding Total Sale. To find the total sale when the net proceeds and the commission rate are known, find the percent the net proceeds is of total sale by subtracting the commission rate from 100%, and dividing the net proceeds by this difference (percent in decimal form). See Figure 6-19. For example, to find the total sale of a product with net proceeds of $200.00 with an 8% commission rate, subtract 8% from 100% (100% − 8% = 92%). Divide 200 by .92 (200 ÷ .92 = 217.3913 = **$217.39**).

FINDING TOTAL SALE

❶

$$\overbrace{100\% - 25\%}^{} = 75\%$$
$$\underbrace{\$2550.00 \div .75}_{} = \mathbf{\$3400.00}$$

❷

❶ SUBTRACT COMMISSION RATE FROM 100%

❷ DIVIDE NET PROCEEDS BY DIFFERENCE (DECIMAL FORM)

Figure 6-19. To find total sale, subtract the commission rate from 100% and divide the net proceeds by the difference (percent in decimal form).

EXAMPLE
Finding Total Sale

1. Find the total sale of a product with net proceeds of $1200.00 with a 4% commission rate.

❶ { **100% − 4% = 96%**

❷ { **$1200.00 ÷ .96 = $1250.00**

❶ SUBTRACT COMMISSION RATE FROM 100%
❷ DIVIDE NET PROCEEDS BY DIFFERENCE (DECIMAL FORM)

① Subtract 4% from 100% (100 − 4 = 96%). ② Divide 1200 by .96 (1200 ÷ .96 = **$1250.00**).

Finding Commission Rate. To find the commission rate when the total sale and net proceeds are known, divide the net proceeds by the total sale, multiply by 100, and subtract the product from 100%. See Figure 6-20.

FINDING COMMISSION RATE

❶ { $1200.00 ÷ $1500.00

❷ { = .8 x 100 = 80%

❸ { 100% – 80% = 20%

❶ DIVIDE NET PROCEEDS BY TOTAL SALE

❷ MULTIPLY QUOTIENT BY 100

❸ SUBTRACT PRODUCT FROM 100%

Figure 6-20. To find commission rate, divide the net proceeds by the total sale, multiply by 100, and subtract the product from 100%.

For example, to find the commission rate on a total sale of $50.00, and net proceeds of $35.00, divide 35 by 50 (35 ÷ 50 = .7). Multiply .7 by 100 (.7 × 100 = 70%). Subtract 70% from 100% (100% – 70% = **30%**).

 EXAMPLE

Finding Commission Rate

1. Find the commission rate on a total sale of $7500.00 and net proceeds of $6750.00.

❶ { $6750.00 ÷ $7500.00

❷ { = .9 x 100 = 90%

❸ { 100% – 90% = 10%

❶ DIVIDE TOTAL SALE BY NET PROCEEDS

❷ MULTIPLY QUOTIENT BY 100

❸ SUBTRACT PRODUCT FROM 100%

① Divide 6750 by 7500 (6750 ÷ 7500 = .9). ② Multiply .9 by 100 (.9 × 100 = 90%). ③ Subtract 90% from 100% (100% – 90% = **10%**).

PRACTICE PROBLEMS

Commission

1. Find the net proceeds for a $25.00 sale for which a $1.25 commission is paid.
2. Find the commission rate on the Used Car.

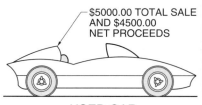

$5000.00 TOTAL SALE AND $4500.00 NET PROCEEDS

USED CAR

3. Find the commission made on a $2250.00 sale with an 8% commission rate.
4. What is the commission rate on a $9000.00 sale with net proceeds of $8460.00?

5. What is the selling price of the House?

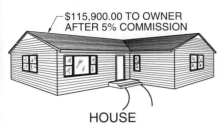

$115,900.00 TO OWNER
AFTER 5% COMMISSION

HOUSE

6. An agent whose rate of commission is 5% gives $3800.00 to a merchant after deducting the commission. What is the total sale?

Tax

A *tax* is a charge paid on income, products, and services. Taxes are a means of raising money to pay the expenses of city, county, state, and federal governments. For example, taxes are paid on salaries, royalties, savings, personal property, automobiles, amusements, etc. *Tax rate* is the percentage of the selling prices, value of property, etc. *Total sale* is the marked price plus the tax.

Finding Tax. To find the tax paid on a product, multiply the marked price by the tax rate (decimal form) to find the tax, and add the tax to the marked price. See Figure 6-21. For example, to find the tax paid on a $25.00 product with a 7% tax rate, multiply 25 by .07 (25 × .07 = **$1.75**).

FINDING TAX

❶ { **$10.00 x .07 = $.70**

❶ MULTIPLY MARKED PRICE BY TAX RATE (DECIMAL FORM)

Figure 6-21. To find tax, multiply the marked price by the tax rate in decimal form.

EXAMPLE

Finding Tax

1. Find the tax paid on a $5200.00 product with an 8% tax rate.

❶ { **$5200.00 x .08 = $416.00**

❶ MULTIPLY MARKED PRICE BY TAX RATE (DECIMAL FORM)

① Multiply 5200 by .08 (5200 × .08 = **$416.00**).

Finding Total Sale. To find the total sale of a taxed product, find the tax and add the marked price. See Figure 6-22. For example, to find the total sale of a $25.00 product with a 7% tax rate, multiply 25 by .07 (25 × .07 = 1.75). Add 25 to 1.75 (25 + 1.75 = **$26.75**).

FINDING TOTAL SALE

❶ { **$10.00 + $.70 = $10.70**

❶ ADD TAX TO MARKED PRICE

Figure 6-22. To find total sale, find the tax and add the marked price.

EXAMPLE

Finding Total Sale

1. Find the total sale of a $56.00 product with a 5% tax rate.

➊ { **$56.00 x .05 = 2.80** ⌐TAX

➋ { **56.00 + 2.80 = $58.80**

➊ MULTIPLY MARKED PRICE BY TAX RATE (DECIMAL RATE)
➋ ADD TAX TO MARKED PRICE

① Multiply 56 by .05 to find the tax (56 × .05 = 2.80). ② Add 56 to 2.80 (56 + 2.80 = **$58.80**).

PRACTICE PROBLEMS

Tax

1. A customer purchases three cans of hair spray at $1.89 per can. The sales tax is 6%. What is the total sale?

2. What is the tax paid on $250.00 with an 11% tax rate?

3. Find the total sale of a $134,000.00 product with a 6% tax rate.

4. To finish a job, an electrician purchases $223.00 of electrical supplies with a tax rate of 7%. What is the total sale of the supplies?

5. What is the total sale of a television that costs $439.00 with a tax rate of 5%?

6. At 7%, how much tax is paid on a $140.00 purchase?

7. Find the tax paid on the Table Saw.

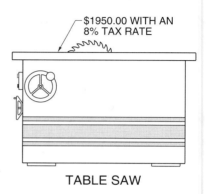

⌐$1950.00 WITH AN 8% TAX RATE

TABLE SAW

8. Find the tax paid on a $60.00 purchase. The tax rate is 4 ½%.

9. What is the tax paid on the screwdriver set at a 6% tax rate?

MADE IN USA

$19.95

8-PIECE SCREWDRIVER SET

10. A 16 oz claw hammer is purchased at $14.50. The tax rate is 6%. What is the total cost?

Name _____ Date _____

True-False and Completion

T F 1. A fractional hundredth is a common fraction having a numerator of 100 and a denominator equal to the percent.

_____ 2. A(n) _____ is a charge paid on items, and is a means of raising money to pay the expense of city, county, state, and federal governments.

T F 3. All fractions and decimals have equivalent percentages.

T F 4. Net price is the price at which products are originally intended to be sold.

_____ 5. A(n) _____ is a method of expressing a part of a whole number in terms of hundredths.

_____ 6. The _____ is the amount of a loan (money borrowed).

T F 7. The discount is the price at which products are originally intended to be sold.

_____ 8. The _____ is the percent of the principal paid back.

_____ 9. The _____ rate is the percentage used to figure commissions.

T F 10. A percent is one part of one hundred parts.

Calculations

_____ 1. Change 84% to a decimal.

_____ 2. Change 14% to a decimal.

_____ 3. Change 71% to a decimal.

_____ **4.** Change 2% to a fractional hundredth.

_____ **5.** Change 35% to a fractional hundredth.

_____ **6.** Change ¾ to a percent.

_____ **7.** Thirty-five percent of what number is 350?

_____ **8.** Forty percent of what number is 3600?

_____ **9.** What percent of 300 is 20?

_____ **10.** What percent of 80 is 16?

_____ **11.** Change ¼ to a percent.

_____ **12.** Change ⅗ to a percent.

_____ **13.** Round .376 to decimal hundredths.

_____ **14.** Round 3.501 to decimal hundredths.

_____ **15.** Round 1.964 to decimal hundredths.

_____ **16.** A furniture store advertises "⅕ off on all stock." What is the percent discount?

_____ **17.** If there is a 3% tax on all retail sales, and a television set costs $450.00 retail, how much tax is added?

_____ **18.** If 2000 lb of iron ore are actually 70% iron, how many pounds of iron are there in the iron ore?

19. A $500.00 bond pays 5% per year. How much interest is earned the first year?

20. An electrician bought 800′ of copper conductor, and 55% of it is used for a wiring job. How many feet are used?

21. How much is saved when buying a tire at a reduction of 30% if the regular price is $40.00?

22. When water freezes, it expands 10% in volume. How many cubic inches (cu in.) of ice does 947 cu in. of water make when frozen?

23. Because of one worn cavity in a compression molding die, a factory lost 10 out of every 40 plastic components. What percent of components is lost?

24. The original price of an automobile is increased 15% ($2250.00). What is the original price?

25. Find the interest on $2500.00 at $5\frac{1}{2}$% for four years.

26. An agent purchased 5000 bushels of corn at $1.30 a bushel. What is the commission at 3%?

27. A grinder has a marked price of $159.00, and a discount price of $119.25. What is the percent discount?

_____ **28.** A 1725 rpm motor has a synchronous speed of 1800 rpm. *Slip* is the percentage difference between actual speed and synchronous speed. What is the slip of the motor?

_____ **29.** A casting, when first poured, is 20.25 centimeters (cm) long. After cooling, the casting shrinks .25 cm. What is the percentage of shrinkage?

_____ **30.** A pump discharges 4500 liters (l) of water per hour running at 75% of its capacity. How many liters per hour does the pump discharge at full capacity?

_____ **31.** A D sheet of drawing paper has a width of 22″. An E sheet of drawing paper has a width of 34″. An E sheet is what percentage wider than a D sheet?

_____ **32.** An 8″ fillet weld is increased in length by 50%. What is the total length of the fillet weld?

_____ **33.** A .750″ drill is what percentage larger than a .500″ drill?

_____ **34.** A 3.00″ block is milled to 2.75″. What percentage is removed by milling?

_____ **35.** Voltage readings on a 230 V motor fluctuate between 215 V and 230 V. What is the percentage of fluctuation?

Name _____ Date _____

_____ 1. Change $\frac{4}{5}$ to a percent.

_____ 2. Eight percent of what number is 240?

_____ 3. Change 14% to a fractional hundredth.

_____ 4. How much tax is paid on $125.00 with a 5% tax rate?

_____ 5. Thirty-two percent of what number is 384?

_____ 6. Change $\frac{5}{7}$ to a percent.

_____ 7. Change 40% to a decimal.

_____ 8. Forty-five percent of what number is 360?

_____ 9. What percent of 130 is 19.50?

_____ 10. Find the commission for $380.00 at an 8% commission rate.

_____ 11. A contractor estimated the cost of a building to be $122,000.00. If 15% is added for profit and overhead, find the total cost.

_____ 12. If the marked price of an article is $60.00, and the discount is $20.00, find the percent discount.

_____ 13. What is the interest on $1250.00 at 8% for three years?

_____ **14.** A boring machine has a speed of 1200 revolutions per minute (rpm). In order to increase the output, the speed is raised $8\frac{1}{5}\%$. What is the new speed?

_____ **15.** A circuit carries 20 A and is protected by a fuse rated at 125% of the circuit ampacity. What size fuse is required?

_____ **16.** If the marked price of an item is $50.00, and the sale price is $37.00, what is the percent discount?

_____ **17.** A steel car is loaded to 70% capacity with 98,000 lb of coal. What is the full capacity of the car?

_____ **18.** If 20 problems are correct on an examination consisting of 25 problems, what is the percentage correct?

_____ **19.** An engine has an output of 50 HP, which is 80% of the input. What is the input?

_____ **20.** A plumber estimates that 500′ of pipe is needed for a job. If 12% is added for waste, how much pipe is ordered?

MEASUREMENTS

The two most commonly used systems of measurement are the English system and the metric system. Conversion factors are used to change measurements from one system to the other. The reduction of denominate numbers changes the units of measure without changing the values.

DENOMINATE NUMBERS

An *abstract number* is any number used by itself. For example, 1, 1½, and 1.5 are all abstract numbers. A *concrete number* is any number used with a name of something or a unit of measure. See Figure 7-1.

A *unit of measure* is a standard by which a quantity such as length, area, capacity, or weight is measured. A *denominate number* is a concrete number that has one unit of measure. A *compound denominate number* is a denominate number that has more than one unit of measure. For example, 6'-8"; 2 hr, 40 min; and 7 lb, 12 oz are compound denominate numbers.

The reduction of denominate numbers is the process of changing denominate numbers from one unit of measure to another without changing the value.

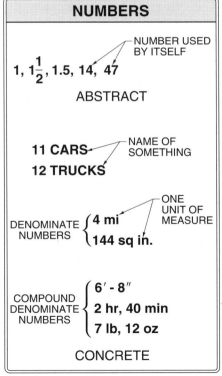

Figure 7-1. Numbers may be abstract or concrete.

The change may be from a higher to a lower denomination or from a lower to a higher denomination. For example, 15″ = 1′-3″.

The common measurement systems are the British (U.S.) system and the SI metric system (International System of Units). Arabic numbers are used with these measurement systems. Conversions can be made from one system to the other.

British (U.S.) System

The British (U.S.) system, primarily used in the United States, is also known as the English system. This system uses the inch (in. or ″), foot (ft or ′), pint (pt), quart (qt), gallon (gal.), ounce (oz), pound (lb), etc. as basic units of measure. See Figure 7-2. See Appendix. In the English system, a denominate number can be changed to different denominations without changing the value.

Changing to Lower Denominations.

To change a compound denominate number to a number of lower denomination, multiply the quantity of highest denomination by the number of units of next lower denomination equal to one unit of the quantity of higher denomination. Add this product to the number of lower denomination in consideration.

Proceed into lower terms in this manner until the required denomination is reached. For example, to change 3′-5″ to inches, multiply 3 by 12, the number of inches in a foot (3 × 12 = 36). Add 36 and 5, the remaining number of inches (36 + 5 = **41″**).

EXAMPLES

Changing to Lower Denominations

1. Find the number of feet in 6 mi, 116 yd, 2 ft.

❶ ⎰ HIGHEST DENOMINATION — PRODUCT
  ⎱ 6 x 1760 = 10,560
  UNITS OF NEXT LOWER DENOMINATION

❷ ⎰ 10,560 + 116 = 10,676

❸ ⎰ 10,676 x 3 = 32,028
  ⎱ 32,028 + 2 = **32,030′**

❶ MULTIPLY HIGHEST DENOMINATION BY UNITS OF NEXT LOWER DENOMINATION
❷ ADD PRODUCT TO NUMBER OF UNITS IN NEXT LOWER DENOMINATION
❸ REPEAT STEPS 1 AND 2 AS REQUIRED

① Multiply 6 by 1760, the number of yards in a mile (6 × 1760 = 10,560). ② Add 10,560 and 116 (10,560 + 116 = 10,676). ③ Multiply 10,676 by 3, the number of feet in a yard (10,676 × 3 = 32,028). Add 32,028 and 2 (32,028 + 2 = **32,030′**).

ENGLISH SYSTEM				
		Unit	**Abbr**	**Equivalents**
LENGTH		mile	mi	5280′, 320 rd, 1760 yd
		rod	rd	5.50 yd, 16.5′
		yard	yd	3′, 36″
		foot	ft *or* ′	12″, .333 yd
		inch	in. *or* ″	.083′, .028 yd
AREA		square mile	sq mi *or* mi²	640 A, 102,400 sq rd
		acre	A	4840 sq yd, 43,560 sq ft
		square rod	sq rd *or* rd²	30.25 sq yd, .00625 A
		square yard	sq yd *or* yd²	1296 sq in., 9 sq ft
		square foot	sq ft *or* ft²	144 sq in., .111 sq yd
		square inch	sq in. *or* in²	.0069 sq ft, .00077 sq yd
VOLUME		cubic yard	cu yd *or* yd³	27 cu ft, 46,656 cu in.
		cubic foot	cu ft *or* ft³	1728 cu in., .0370 cu yd
		cubic inch	cu in. *or* in³	.00058 cu ft, .000021 cu yd
CAPACITY	*U.S. liquid measure*	gallon	gal.	4 qt (231 cu in.)
		quart	qt	2 pt (57.75 cu in.)
		pint	pt	4 gi (28.875 cu in.)
		gill	gi	4 fl oz (7.219 cu in.)
		fluidounce	fl oz	8 fl dr (1.805 cu in.)
		fluidram	fl dr	60 min (.226 cu in.)
		minim	min	⅙ fl dr (.003760 cu in.)
	U.S. dry measure	bushel	bu	4 pk (2150.42 cu in.)
		peck	pk	8 qt (537.605 cu in.)
		quart	qt	2 pt (67.201 cu in.)
		pint	pt	½ qt (33.600 cu in.)
	British imperial liquid and dry measure	bushel	bu	4 pk (2219.36 cu in.)
		peck	pk	2 gal. (554.84 cu in.)
		gallon	gal.	4 qt (277.420 cu in.)
		quart	qt	2 pt (69.355 cu in.)
		pint	pt	4 gi (34.678 cu in.)
		gill	gi	5 fl oz (8.669 cu in.)
		fluidounce	fl oz	8 fl dr (1.7339 cu in.)
		fluidram	fl dr	60 min (.216734 cu in.)
		minim	min	1⁄60 fl dr (.003612 cu in.)
MASS AND WEIGHT	*avoirdupois*	ton		2000 lb
		short ton	t	2000 lb
		long ton		2240 lb
		pound	lb *or* #	16 oz, 7000 gr
		ounce	oz	16 dr, 437.5 gr
		dram	dr	27.344 gr, .0625 oz
		grain	gr	.037 dr, .002286 oz
	troy	pound	lb	12 oz, 240 dwt, 5760 gr
		ounce	oz	20 dwt, 480 gr
		pennyweight	dwt *or* pwt	24 gr, .05 oz
		grain	gr	.042 dwt, .002083 oz
	apothecaries'	pound	lb ap	12 oz, 5760 gr
		ounce	oz ap	8 dr ap, 480 gr
		dram	dr ap	3 s ap, 60 gr
		scruple	s ap	20 gr, .333 dr ap
		grain	gr	.05 s, .002083 oz, .0166 dr ap

Figure 7-2. The English system is based on inches, feet, pints, quarts, gallons, ounces, pounds, etc.

2. Find the number of pints in 7 gal., 3 qt, 1pt.

❶$\Big\{$ 7 x 4 = 28

❷$\Big\{$ 28 + 3 = 31

❸$\Big\{$ 31 x 2 = 62
62 + 1 = **63 pt**

❶ MULTIPLY HIGHEST DENOMINATION BY UNITS OF NEXT LOWER DENOMINATION

❷ ADD PRODUCT TO NUMBER OF UNITS IN NEXT LOWER DENOMINATION

❸ REPEAT STEPS 1 AND 2 AS REQUIRED

① Multiply 7 by 4, the number of quarts in a gallon (7 × 4 = 28). ② Add 28 and 3 (28 + 3 = 31). ③ Multiply 31 by 2, the number of pints in a quart (31 × 2 = 62 pt). Add 62 and 1 (62 + 1 = **63 pt**).

Changing to Higher Denominations. To change a denominate number to a compound number of higher denomination, divide the given number by the number of units contained in one unit of the next higher denomination. Set aside any remainder and give it the name of the units of the dividend.

Divide the quotient obtained by the number of units contained in one unit of the next higher denomination, and set aside the remainder with its proper name. Proceed until the required denomination is reached. The answer is the last quotient and the several properly named remainders.

For example, to change 765 pt to gallons (higher denomination), change pints to quarts by dividing 765 by 2 (765 ÷ 2 = 382 and a remainder of 1 pt). Divide 382 by 4 (382 ÷ 4 = 95 and a remainder of 2 qt). The quantity of 765 pt = **95 gal., 2 qt, 1 pt**.

 EXAMPLES

Changing to Higher Denominations

1. Change 7,065,000 sq in. to acres.

UNITS OF NEXT HIGHER DENOMINATION ―
┌GIVEN NUMBER ＼ ┌REMAINDER

❶$\Big\{$7,065,000 ÷ 144 = 49,062.5 =
└QUOTIENT

49,062 sq ft, 72 sq in.

❷$\Big\{$49,062 ÷ 9 = 5451 sq yd, 3 sq ft

❸$\Big\{$5451 ÷ 4840 = **1 A, 611 sq yd, 3 sq ft, 72 sq in.**

❶ DIVIDE GIVEN NUMBER BY UNITS OF NEXT HIGHER DENOMINATION. SET ASIDE ANY REMAINDERS

❷ DIVIDE QUOTIENT BY UNITS OF NEXT HIGHER DENOMINATION. SET ASIDE ANY REMAINDERS

❸ REPEAT STEP 2 AS REQUIRED

① Divide 7,065,000 by 144, the number of square inches in a square foot (7,065,000 ÷ 144 = 49,062.5

sq ft = 49,062 sq ft and a remainder of 72 sq in.). ② Divide 49,062 by 9, the number of square feet in a square yard (49,062 ÷ 9 = 5451 and a remainder of 3 sq ft). ③ Divide 5451 by 4840, the number of square yards per acre (5451 ÷ 4840 = 1 and a remainder of 611 sq yd). The area in acres of 7,065,000 sq in. = **1 A, 611 sq yd, 3 sq ft, 72 sq in.**

2. Change 2000 cu in. to cubic feet.

❶{ **2000 ÷ 1728**
= 1 cu ft, 272 cu in.

❶ DIVIDE GIVEN NUMBER BY UNITS OF NEXT HIGHER DENOMINATION. SET ASIDE ANY REMAINDERS

① Divide 2000 by 1728, the number of cubic inches in a cubic foot (2000 ÷ 1728 = **1 cu ft, 272 cu in.**).

Adding and Subtracting. To add or subtract denominate numbers, one of two methods is used. The first method is to change the given numbers to the lowest denomination in the problem, perform the required operation, change the result to higher denominations if necessary, and add.

For example, to add 3 mi, 356 yd, 2′ and 1 mi, 29 yd, 1′; multiply 3 and 1 by 5280, the number of feet in a mile (3 × 5280 = 15,840 and 1 × 5280 = 5280). Multiply 356 and

29 by 3, the number of feet in a yard (356 × 3 = 1068 and 29 × 3 = 87). Add 15,840, 5280, 1068, 87, 2, and 1 (15,840 + 5280 + 1068 + 87 + 2 + 1 = 22,278′).

Change feet to yards by dividing 22,278 by 3 (22,278 ÷ 3 = 7426 yd). Change yards to miles by dividing 7426 by 1760 (7426 ÷ 1760 = 4 and a remainder of 386). The distance of 3 mi, 356 yd, 2′ + 1 mi, 29 yd, 1′ = **4 mi, 386 yd**.

The second method of adding and subtracting denominate numbers is ordinarily more appropriate because it is similar to adding and subtracting abstract numbers. The second method is to perform the operations on the given numbers, change denominations as required during the operations, and add.

For example, to add 3 mi, 356 yd, 2′ and 1 mi, 29 yd, 1′; add the miles (3 + 1 = 4 mi). Add the yards (356 + 29 = 385 yd). Add the feet (2 + 1 = 3′). Change the 3′ to 1 yd and add it to the yards (385 + 1 = 386 yd). No more changes in denominations can be made. The distance of 3 mi, 356 yd, 2′ + 1 mi, 29 yd, 1′ = **4 mi, 386 yd**.

 EXAMPLES

Adding and Subtracting

1. Add 15 mi, 45 yd, 2′ and 3 mi, 89 yd, 1′.

GIVEN
NUMBERS

❶ $\begin{cases} 15 + 3 = 18 \text{ mi} \\ 45 + 89 = 134 \text{ yd} \\ 2 + 1 = 3' = 1 \text{ yd} \\ 134 + 1 = 135 \text{ yd} \\ = 18 \text{ mi, 135 yd} \end{cases}$

❶ PERFORM GIVEN OPERATION

① Add the miles (15 + 3 = 18 mi). Add the yards (45 + 89 = 134 yd). Add the feet (2 + 1 = 3'). Change the 3' to 1 yd. Add 1 to the yards (134 + 1 = 135 yd). No more changes in denominations can be made. The length of 15 mi, 45 yd, 2' + 3 mi, 89 yd, 1' = **18 mi, 135 yd**.

2. Subtract 1 gal., 3 qt, 1 pt from 6 gal., 3 qt, 2 pt.

❶ $\begin{cases} 6 - 1 = 5 \text{ gal.} \\ 3 - 3 = 0 \text{ qt} \\ 2 - 1 = 1 \text{ pt} \\ = 5 \text{ gal., 1 pt} \end{cases}$

❶ PERFORM GIVEN OPERATION

① Subtract the gallons (6 − 1 = 5 gal.). Subtract the quarts (3 − 3 = 0 qt). Subtract the pints (2 − 1 = 1 pt). The volume of 6 gal., 3 qt, 2 pt − 1 gal., 3 qt, 1 pt = **5 gal., 1 pt**.

Multiplying. To multiply denominate numbers, multiply the same as abstract numbers, change the denominations as required, and add. For example, to multiply 5 gal., 3 qt, 1 pt by 5, multiply 5 by 5 (5 ×

5 = 25 gal.). Multiply 3 by 5 (3 × 5 = 15 qt). Multiply 1 by 5 (1 × 5 = 5 pt). Change 5 pints to quarts by dividing 5 by 2 (5 ÷ 2 = 2 qt and a remainder of 1 pt). Add 2 to 15 (15 + 2 = 17 qt). Change 17 qt to gallons by dividing 17 by 4 (17 ÷ 4 = 4 gal. and a remainder of 1 qt). Add 4 to 25 (25 + 4 = 29 gal.). The volume of 5 gal., 3 qt, 1 pt × 5 = **29 gal., 1 qt, 1 pt**.

EXAMPLES

Multiplying

1. Multiply 15 yd, 2', 8" by 4.

❶ $\begin{cases} 15 \times 4 = 60 \text{ yd} \\ 2 \times 4 = 8' \\ 8 \times 4 = 32'' \end{cases}$

❷ $\begin{cases} 32 \div 12 = 2', 8'' \\ 8 \div 3 = 2 \text{ yd, 2'} \\ 60 \text{ yd} + 2 \text{ yd, 2'} + 2', 8'' \end{cases}$
= **63 yd, 1', 8"**

❶ MULTIPLY SAME AS ABSTRACT NUMBERS
❷ CHANGE DENOMINATOR AS REQUIRED AND ADD

① Multiply 15 by 4 (15 × 4 = 60 yd). Multiply 2 by 4 (2 × 4 = 8'). Multiply 8 by 4 (8 × 4 = 32"). ② Change 32" to feet by dividing by 12 (32 ÷ 12 = 2' and a remainder of 8"). Change 8' to yards by dividing 8 by 3 (8 ÷ 3 = 2 yd and a remainder of 2'). The length of 15 yd, 2', 8" × 4 = **63 yd, 1', 8"**.

2. Multiply 12 cu yd, 24 cu ft by 3.

❶ $\begin{cases} \textbf{12 x 3 = 36 cu yd} \\ \textbf{24 x 3 = 72 cu ft} \end{cases}$

❷ $\begin{cases} \textbf{72 ÷ 27 = 2 cu yd, 18 cu ft} \\ \textbf{36 cu yd, + 2 cu yd + 18 cu ft} \end{cases}$
= 38 cu yd, 18 cu ft

❶ MULTIPLY SAME AS
ABSTRACT NUMBERS
❷ CHANGE DENOMINATION
AS REQUIRED AND ADD

① Multiply 12 by 3 (12 × 3 = 36 cu yd). Multiply 24 by 3 (24 × 3 = 72 cu ft). Change 72 cu ft to cubic yards by dividing by 27 (72 ÷ 27 = 2 cu yd and a remainder of 18 cu ft). The volume of 12 cu yd, 24 cu ft × 3 = **38 cu yd, 18 cu ft**.

EXAMPLE
Dividing

1. Divide 9 yd, 2 ft, 3 in. by 3.

❶ $\begin{cases} \textbf{9 ÷ 3 = 3 yd} \end{cases}$

❷ $\begin{cases} \textbf{2 x 12 = 24''} \\ \textbf{24 + 3 = 27''} \\ \textbf{27 ÷ 3 = 9''} \end{cases}$
= 3 yd, 0', 9''

❶ DIVIDE SAME AS
ABSTRACT NUMBERS
❷ CHANGE DENOMINATION
AS REQUIRED AND ADD

① Divide 9 by 3 (9 ÷ 3 = 3 yd). ② Multiply 2 by 12 (2 × 12 = 24''). Add 24 and 3 (24 + 3 = 27''). Divide 27 by 3 (27 ÷ 3 = 9''). The length of 9 yd, 2 ft, 3 in. ÷ 3 = **3 yd, 0', 9''**.

Dividing. To divide denominate numbers, divide the same as abstract numbers, change the denominations as required, and add. For example, to divide 15 gal., 2 qt, 2 pt by 2, divide 15 by 2 (15 ÷ 2 = 7 and a remainder of 1 gal.). Change 1 gal. to quarts by multiplying by 4 (1 × 4 = 4 qt). Add 4 to 2 (2 + 4 = 6 qt). Divide 6 by 2 (6 ÷ 2 = 3 qt). Divide 2 by 2 (2 ÷ 2 = 1 pt). The volume of 15 gal., 2 qt, 2 pt ÷ 2 = **7 gal., 3 qt, 1 pt**.

PRACTICE PROBLEMS
British (U.S.) System

1. Change 244 sq yd to square inches.
2. Change 25 t, 70 lb to pounds.
3. Change 2 mi, 192 rd, 2 yd, to feet.
4. Find the sum of 10 yd, 2', 10'' + 15 yd, 1', 9'' + 8 yd, 2', 7'' + 18 yd, 1', 11'' + 16 yd, 2', 8''.
5. Multiply 4 yd, 2', 8'' by 5.
6. Express 1507 pt in gallons, quarts, and pints.

7. Express 17,000 ft in miles, yards, and feet.
8. Subtract 4 yd, 2', 10" from 15 yd, 2', 7".
9. How much oil is in the two Oil Drums?

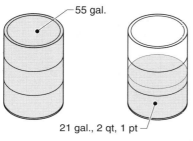

55 gal.

21 gal., 2 qt, 1 pt

OIL DRUMS

10. How many cubic feet of sand can be moved in five trips with the wheelbarrow filled to maximum capacity each trip?

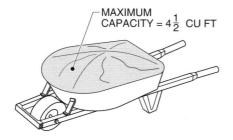

MAXIMUM
CAPACITY = $4\frac{1}{2}$ CU FT

11. How many square feet does a ½ A building lot contain?
12. The distance from a baseball field pitching rubber to home plate is 60'-6". What is this distance in inches?
13. How many rods are in a mile?
14. How many ounces are in a ton?

15. Which is the larger area, 1 sq yd or 1 sq rd?
16. What is the total perimeter length of Lot 312?

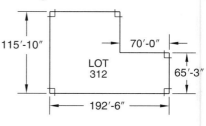

115'-10"   70'-0"

LOT 312

65'-3"

192'-6"

Metric System

The metric system is the most common measurement system used in the world. This system is based on the meter (m), liter (l), and gram (g). See Figure 7-3. See Appendix. Prefixes are used in the metric system to represent multipliers. For example, the prefix *kilo* (*k*) has a prefix equivalent of 1000, so 1 kilometer (km) = 1000 m = 1 km.

To change from a quantity to another prefix, multiply the quantity by the number of units that equals one of the original metric units (conversion factor). For example, to change 544 m to kilometers (1 m = $^1/_{1000}$ k), multiply 544 by .001 (544 × .001 = **.544 km**). To change 35 l to centiliters (1 cl = 100 l), multiply 35 by 100 (35 × 100 = **3500 cl**).

METRIC SYSTEM			
	Unit	Abbr	Number of Base Units
LENGTH	kilometer	km	1000
	hectometer	hm	100
	dekameter	dam	10
	*meter	m	1
	decimeter	dm	.1
	centimeter	cm	.01
	millimeter	mm	.001
AREA	square kilometer	sq km *or* km^2	1,000,000
	hectare	ha	10,000
	are	a	100
	square centimeter	sq cm *or* cm^2	.0001
VOLUME	cubic centimeter	cu cm, cm^3, *or* cc	.000001
	cubic decimeter	dm^3	.001
	*cubic meter	m^3	1
CAPACITY	kiloliter	kl	1000
	hectoliter	hl	100
	dekaliter	dal	10
	*liter	l	1
	cubic decimeter	dm^3	1
	deciliter	dl	.10
	centiliter	cl	.01
	milliliter	ml	.001
MASS AND WEIGHT	metric ton	t	1,000,000
	kilogram	kg	1000
	hectogram	hg	100
	dekagram	dag	10
	*gram	g	1
	decigram	dg	.10
	centigram	cg	.01
	milligram	mg	.001

*Base Units

Figure 7-3. The metric system is based on the meter, liter, and gram.

It may be necessary to change a quantity to a base unit before changing to a new prefix. For example, to change 2000 mg to decagrams, change 2000 mg to grams (base unit) by multiplying 2000 by .001 (2000 × .001 = 2). Change grams to decagrams by multiplying 2 by .1 (2 × .1 = **.2 dg**).

Additionally, a conversion table may be used to change a metric quantity to another prefix. For example, to change from liters to kiloliters, move the decimal point three places to the right. See Figure 7-4. See Appendix.

EXAMPLES
Metric System

1. Change 134 kg to grams.

❶ { 134 × 1000 = **134,000 g**
⌞CONVERSION FACTOR

❶ MULTIPLY BY CONVERSION FACTOR

① Multiply 134 by the conversion factor of 1000 (134 × 1000 = **134,000 g**).

2. Change 14 mm to meters.

❶ { 14 × .000001 = **.000014 m**

❶ MULTIPLY BY CONVERSION FACTOR

① Multiply 14 by the conversion factor .000001 (14 × .000001 = **.000014 m**).

3. Change 50,000 ml to kiloliters.

❶ { 50,000 × .001 = 50 l
{ 50 × .001 = **.05 kl**

❶ MULTIPLY BY CONVERSION FACTOR

① Multiply 50,000 by the conversion factor .001 (50,000 × .001 = 50 l). Change liters to kiloliters by multiplying 50 by the conversion factor .001 (50 × .001 = **.05 kl**).

PRACTICE PROBLEMS
Metric System

1. Change .0025 m to millimeters.
2. Change 24,768 g to kilograms. *(3 places)*
3. Change 375 mm to centimeters.
4. Change 555 cm to millimeters.
5. Change 5310 cm to meters.
6. Change 344 l to dekaliters.
7. Change 3 kl to hectoliters.
8. What is the diameter of each circle in millimeters?

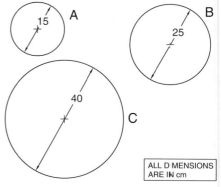

ALL DIMENSIONS ARE IN cm

MOVE DECIMAL POINT

544 m = ? km THREE PLACES TO LEFT

544 x .001 = .544 km **OR** **.544 m = ? km**

CONVERSION **544 m = .544 km**
FACTOR

MOVE DECIMAL POINT

35 l = ? cl TWO PLACES TO RIGHT **35 l = ? cl**

35 x 100 = 3500 cl **OR** **35 l = 3500 cl**

CONVERSION
FACTOR

PREFIXES

Multiples and Submultiples	Prefixes	Symbols	Meaning
$1,000,000,000,000 = 10^{12}$	tera	T	trillion
$1,000,000,000 = 10^{9}$	giga	G	billion
$1,000,000 = 10^{6}$	mega	M	million
$1000 = 10^{3}$	kilo	k	thousand
$100 = 10^{2}$	hecto	h	hundred
$10 = 10^{1}$	deka	d	ten
Unit $1 = 10^{0}$			
$.1 = 10^{-1}$	deci	d	tenth
$.01 = 10^{-2}$	centi	c	hundredth
$.001 = 10^{-3}$	milli	m	thousandth
$.000001 = 10^{-6}$	micro	μ	millionth
$.000000001 = 10^{-9}$	nano	n	billionth
$.000000000001 = 10^{-12}$	pico	p	trillionth

CONVERSION TABLE

Initial Units	Final Units											
	giga	mega	kilo	hecto	deka	base unit	deci	centi	milli	micro	nano	pico
giga		3R	6R	7R	8R	9R	10R	11R	12R	15R	18R	21R
mega	3L		3R	4R	5R	6R	7R	8R	9R	12R	15R	18R
kilo	6L	3L		1R	2R	3R	4R	5R	6R	9R	12R	15R
hecto	7L	4L	1L		1R	2R	3R	4R	5R	8R	11R	14R
deka	8L	5L	2L	1L		1R	2R	3R	4R	7R	10R	13R
base unit	9L	6L	3L	2L	1L		1R	2R	3R	6R	9R	12R
deci	10L	7L	4L	3L	2L	1L		1R	2R	5R	8R	11R
centi	11L	8L	5L	4L	3L	2L	1L		1R	4R	7R	10R
milli	12L	9L	6L	5L	4L	3L	2L	1L		3R	6R	9R
micro	15L	12L	9L	8L	7L	6L	5L	4L	3L		3R	6R
nano	18L	15L	12L	11L	10L	9L	8L	7L	6L	3L		3R
pico	21L	18L	15L	14L	13L	12L	11L	10L	9L	6L	3L	

R = Move the decimal point to the right.   L = Move the decimal point to the left.

Figure 7-4. To change a metric quantity to another prefix, multiply by the number of units that equals the prefix or use the Conversion Table.

MEASURES

Extension is the property of a body by which it occupies space and has one or more of the three dimensions of length, width, and thickness. The three measures of extension are linear measure, square measure (area), and cubic measure (volume).

The English and metric systems have separate units of measure for the three measures of extension. The two systems also have separate measures and scales for capacity, mass, and temperature.

Conversions

Conversions can be performed in all units of measure. English and metric measurements are converted from one system to the other by multiplying by the number of units of one system that equals one unit of measurement of the other. The number of units is the conversion factor. Conversions are performed by applying the appropriate conversion factor. Equivalent Tables are used to convert between systems. See Figure 7-5. See Appendix.

EQUIVALENT TABLE (METRIC TO ENGLISH)

	Unit	British Equivalent		
LENGTH	kilometer	.62 mi		
	hectometer	109.36 yd		
	dekameter	32.81'		
	meter	39.37"		
	decimeter	3.94"		
	centimeter	.39"		
	millimeter	.039"		
AREA	square kilometer	.3861 sq mi		
	hectare	2.47 A		
	are	119.60 sq yd		
	square centimeter	.155 sq in.		
VOLUME	cubic centimeter	.061 cu in.		
	cubic decimeter	61.023 cu in.		
	cubic meter	1.307 cu yd		
		cubic	*dry*	*liquid*
CAPACITY	kiloliter	1.31 cu yd		
	hectoliter	3.53 cu ft	2.84 bu	
	dekaliter	.35 cu ft	1.14 pk	2 64 gal.
	liter	61.02 cu in.	.908 qt	1 057 qt
	cubic decimeter	61.02 cu in.	.908 qt	1 057 qt
	deciliter	6.1 cu in.	.18 pt	.21 pt
	centiliter	.61 cu in.		338 fl oz
	milliliter	.061 cu in.		.27 fl dr
MASS AND WEIGHT	metric ton	1.102 t		
	kilogram	2.2046 lb		
	hectogram	3.527 oz		
	dekagram	.353 oz		
	gram	.035 oz		
	decigram	1.543 gr		
	centigram	.154 gr		
	milligram	.015 gr		

EQUIVALENT TABLE (ENGLISH TO METRIC)			
		Unit	**Metric Equivalent**
LENGTH		mile	1.609 km
		rod	5.029 m
		yard	.9144 m
		foot	30.48 cm
		inch	2.54 cm
AREA		square mile	2.590 k^2
		acre	.405 hectare, 4047 m^2
		square rod	25.293 m^2
		square yard	.836 m^2
		square foot	.093 m^2
		square inch	6.452 cm^2
VOLUME		cubic yard	.765 m^3
		cubic foot	.028 m^3
		cubic inch	16.387 cm^3
CAPACITY	*U.S. liquid measure*	gallon	3.785 l
		quart	.946 l
		pint	.473 l
		gill	118.294 ml
		fluidounce	29.573 ml
		fluidram	3.697 ml
		minim	.061610 ml
	U.S. dry measure	bushel	35.239 l
		peck	8.810 l
		quart	1.101 l
		pint	.551 l
	British imperial liquid and dry measure	bushel	.036 m^3
		peck	.0091 m^3
		gallon	4.546 l
		quart	1.136 l
		pint	568.26 cm^3
		gill	142.066 cm^3
		fluidounce	28.412 cm^3
		fluidram	3.5516 cm^3
		minim	.059194 cm^3
MASS AND WEIGHT	*avoidupois*	short ton	.907 t
		long ton	1.016 t
		pound	.454 kg
		ounce	28.350 g
		dram	1.772 g
		grain	.0648 g
	troy	pound	.373 kg
		ounce	31.103 g
		pennyweight	1.555 g
		grain	.0648 g
	apothecaries'	pound	.373 kg
		ounce	31.103 g
		dram	3.888 g
		scruple	1.296 g
		grain	.0648 g

Figure 7-5. Equivalent Tables are used to convert between systems.

To convert a metric measurement to English, multiply the measurement by the number of English units that equals one of the metric units (conversion factor). For example, to convert 50 mm to inches, multiply 50 by .03937 ($50 \times .03937 = \textbf{1.9685}''$).

To convert an English measurement to metric, multiply the measurement by the number of metric units that equals one of the English units (conversion factor). For example, to convert 3 mi to kilometers, multiply 3 by 1.609 ($3 \times 1.609 = \textbf{4.827 km}$).

EXAMPLES

Conversions

Metric measurements to English measurements.

1. Convert 3 m to inches.

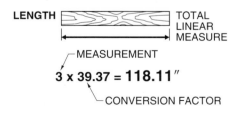

LENGTH — TOTAL LINEAR MEASURE

MEASUREMENT

$3 \times 39.37 = \textbf{118.11}''$

CONVERSION FACTOR

Multiply 3 m by the conversion factor 39.37 ($3 \times 39.37 = \textbf{118.11}''$).

2. Convert 3 ha to acres.

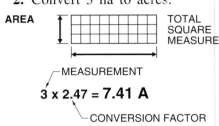

AREA — TOTAL SQUARE MEASURE

MEASUREMENT

$3 \times 2.47 = \textbf{7.41 A}$

CONVERSION FACTOR

Multiply 3 ha by the conversion factor 2.47 ($3 \times 2.47 = \textbf{7.41 A}$).

3. Convert 12 m^3 to cubic yards.

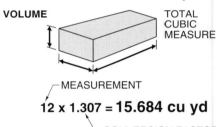

VOLUME — TOTAL CUBIC MEASURE

MEASUREMENT

$12 \times 1.307 = \textbf{15.684 cu yd}$

CONVERSION FACTOR

Multiply 12 m^3 by the conversion factor 1.307 ($12 \times 1.307 = \textbf{15.684 cu yd}$).

4. Convert 6 l to quarts.

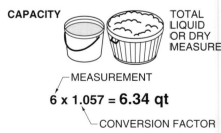

CAPACITY — TOTAL LIQUID OR DRY MEASURE

MEASUREMENT

$6 \times 1.057 = \textbf{6.34 qt}$

CONVERSION FACTOR

Multiply 6 l by the conversion factor 1.057 ($6 \times 1.057 = \textbf{6.34 qt}$).

5. Convert 1500 g to pounds.

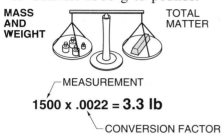

MASS AND WEIGHT · **TOTAL MATTER** · MEASUREMENT · **1500 x .0022 = 3.3 lb** · CONVERSION FACTOR

Multiply 1500 g by the conversion factor .0022 (1500 × .0022 = **3.3 lb**).

English measurements to metric measurements.

1. Convert 3 in. to centimeters.

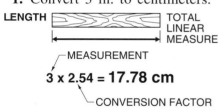

LENGTH · **TOTAL LINEAR MEASURE** · MEASUREMENT · **3 x 2.54 = 17.78 cm** · CONVERSION FACTOR

Multiply 3″ by the conversion factor 2.54 (3 × 2.54 = **17.78″**).

2. Convert 6 sq mi to square kilometers.

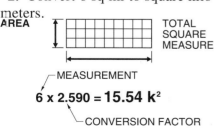

AREA · **TOTAL SQUARE MEASURE** · MEASUREMENT · **6 x 2.590 = 15.54 k²** · CONVERSION FACTOR

Multiply 6 sq mi by the conversion factor 2.590 (6 × 2.590 = **15.54 k²**).

3. Convert 15 cu ft to cubic meters.

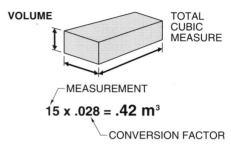

VOLUME · **TOTAL CUBIC MEASURE** · MEASUREMENT · **15 x .028 = .42 m³** · CONVERSION FACTOR

Multiply 15 cu ft by the conversion factor .028 (15 × .028 = **.42 m³**).

4. Convert 3 pt to liters.

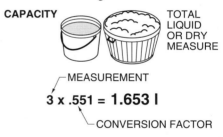

CAPACITY · **TOTAL LIQUID OR DRY MEASURE** · MEASUREMENT · **3 x .551 = 1.653 l** · CONVERSION FACTOR

Multiply 3 pt by the conversion factor .551 (3 × .551 = **1.653 l**).

5. Convert 11 oz to grams.

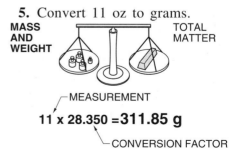

MASS AND WEIGHT · **TOTAL MATTER** · MEASUREMENT · **11 x 28.350 = 311.85 g** · CONVERSION FACTOR

Multiply 11 oz by the conversion factor 28.350 (11 × 28.350 = **311.85 g**).

Linear Measure

Linear measure is the measurement of length. It is used to find the one-dimensional length of an object. It measures distances such as how far, how long, etc. The common units used for linear measure are the inch (") in the English system and the millimeter (mm) in the metric system. See Figure 7-6.

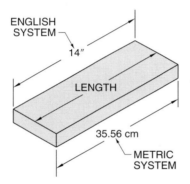

Figure 7-6. Linear measure is the measurement of length.

Square Measure (Area)

Square measure is the measurement of area. It is used to find the two-dimensional size of plane (flat) surfaces. A surface has the two dimensions length and width. Surface measurement is obtained by finding the area. See Figure 7-7.

The *area* of a surface is the number of units of length multiplied by the number of units of width. Area is expressed in square units. For example, the floor of a room has a length and a width. Its area may be expressed in square feet (sq ft), square meters (m^2), etc. The difference between the terms "square feet" and "feet square" is that a 2' square measures 2' on each side. Two square feet (2 sq ft) measures 1.4142' ($\sqrt{2}$) on each side.

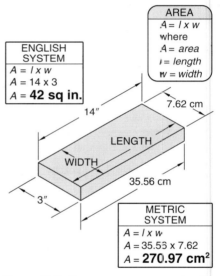

Figure 7-7. Square measure is the measurement of area.

In the English system, the acre is used for the measurement of large areas of land. An *acre* is an area of land containing 43,560 sq ft. It may not be a square unit, although it is equal to the area of a square measuring approximately 209' on each side. Ordinarily, an acre measures a certain number of rods on one side and a certain number of rods on the other, measuring a total surface of 160 sq rd.

In the metric system, the square millimeter and square meter are used for square measure. Large surfaces are measured as ares or hectares. An *are* is an area of land containing 100 m². A *hectare* is an area of land containing 10,000 m².

Cubic Measure (Volume)

Cubic measure is the measurement of volume. It is used to find the three-dimensional size of an object. A solid object has the three dimensions of length, width, and thickness. *Volume* is the contents of a three-dimensional object. See Figure 7-8.

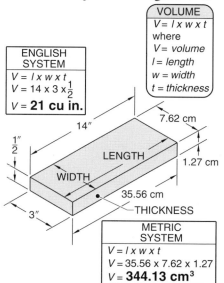

VOLUME
$V = l \times w \times t$
where
V = volume
l = length
w = width
t = thickness

ENGLISH SYSTEM
$V = l \times w \times t$
$V = 14 \times 3 \times \frac{1}{2}$
$V = $ **21 cu in.**

14″

7.62 cm

$\frac{1}{2}″$

LENGTH

1.27 cm

WIDTH

35.56 cm

3″

THICKNESS

METRIC SYSTEM
$V = l \times w \times t$
$V = 35.56 \times 7.62 \times 1.27$
$V = $ **344.13 cm³**

Figure 7-8. Cubic measure is the measurement of volume.

The units of cubic measure are the same as those in linear measure with the prefix "cubic". Length, width, and thickness must be expressed in the same unit of measure. For example, cubic inch (cu in.), cubic yard (cu yd), etc. are used in the English system. Cubic centimeter (cm³), cubic meter (m³) are used in the metric system.

Capacity

Capacity is the measurement of volume that a container can hold. The English system has no unity in the measurements of capacity because it uses of three kinds of measures. In the metric system, there is unity of measurement because only one measurement is used. See Figure 7-9.

The three English system measurements of measure for capacity are the liquid measure, dry measure, and apothecaries' fluid measure. *Liquid measure* is the measurement of liquids, such as water or fuel. It is the most commonly used of the three English system tables of measure for capacity.

The *dry measure* is the measurement of dry items, such as grains, vegetables, etc. The *apothecaries' fluid measure* is the measurement of drugs. It is used for filling medical prescriptions.

In the metric system, the standard unit of measure is based on the liter (l). The liter is larger than the liquid quart but smaller than the dry quart. One liter equals 1.057 liquid quarts and .908 dry quarts.

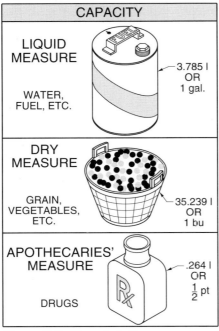

CAPACITY

LIQUID MEASURE

WATER, FUEL, ETC.

3.785 l OR 1 gal.

DRY MEASURE

GRAIN, VEGETABLES, ETC.

35.239 l OR 1 bu

APOTHECARIES' MEASURE

DRUGS

.264 l OR $\frac{1}{2}$ pt

Figure 7-9. Capacity is the measurement of volume that a container can hold.

Capacity and Cubic Measure. Capacity and cubic measures have equivalents that are commonly used to find the capacity of a container that has a certain volume. For example, in the English system, a container with dimensions of $1' \times 1' \times 1'$ holds 7.5 gal. (1 cu ft = 7.5 gal.). A container with dimensions of 10 cm × 10 cm × 10 cm holds one liter (1 l = 1000 cm³).

Mass

Weight is a measure of gravity or the force of the earth's attraction for a body. *Mass* is the measurement of matter contained in an object. The mass of an object never changes, whereas the weight of an object changes in relationship to its distance from earth's center. See Figure 7-10.

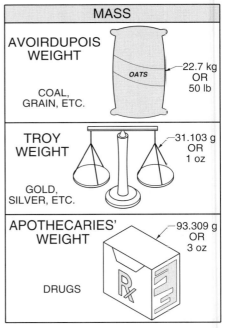

MASS

AVOIRDUPOIS WEIGHT

COAL, GRAIN, ETC.

OATS

22.7 kg OR 50 lb

TROY WEIGHT

GOLD, SILVER, ETC.

31.103 g OR 1 oz

APOTHECARIES' WEIGHT

DRUGS

93.309 g OR 3 oz

Figure 7-10. Mass is the measurement of matter contained in an object.

Weight is measured by scales of various kinds. Spring balances are often used for light objects. Other scales use lever arms with definite marked weights on one side of the lever.

As in the measure of capacity, the English system has three different measurements for weight. These three measurements are avoirdupois weights, troy weights, and apothecaries' weights.

Avoirdupois weight is the measurement of heavy weights. It is used for objects such as coal, grain, livestock, meats, groceries, etc. It is the most commonly used of the three English system measurements for weight.

Troy weight is the measurement of precious materials. It is used for diamonds, gold, silver, etc. *Apothecaries' weight* is the measurement of drugs. It is used for filling prescriptions.

The metric system has one table of measure for mass. It is based on the gram or kilogram. For example, 1 cm³ of water equals 1 g. The gram is small and difficult to work with. The kilogram (kg) is more often used as the base unit.

To change the denomination of a measure within a system, multiply the number of units to be changed by the number of units within the desired measure. For example, to change 10 lb to ounces, multiply 10 by 16, the number of ounces in a pound (10 × 16 = **160 oz**). To change 3500 g to kilograms, multiply 3500 by .001, the number of kilograms in a gram (3500 × .001 = **3.5 kg**).

PRACTICE PROBLEMS

Conversions

1. Convert 5 m to inches.
2. Convert 58 km to miles.
3. Convert 5 sq cm to square inches. *(3 places)*
4. Convert 12 sq km to square miles. *(4 places)*
5. Convert 21 cm³ to cubic inches. *(3 places)*
6. Convert 7 m³ to cubic yards. *(3 places)*
7. Convert 24 kl to cubic yards.
8. Convert 2 l to qt (dry). *(3 places)*
9. Convert 17 kg to lb. *(4 places)*
10. Convert 12 g to ounces.
11. Convert 22 mi to kilometers. *(3 places)*
12. Convert 5 in. to centimeters.
13. Convert 60 A to hectares.
14. Convert 190 sq ft to square meters.
15. Convert 5.5 cu in. to cubic centimeters. *(4 places)*
16. Convert 21 cu ft to cubic meters. *(3 places)*
17. Convert 2 bu to liters. *(3 places)*
18. Convert 2 gal. to liters. *(3 places)*
19. Convert 8 troy oz to grams.

Temperature

Temperature is the measurement of hotness or coldness. Temperatures are measured with thermometers. A *thermometer* is an instrument used for measuring temperature. Celsius

(°C) and Fahrenheit (°F) are the two temperature scales most commonly used for measuring temperature. Conversions can be made from one scale to the other. See Figure 7-11.

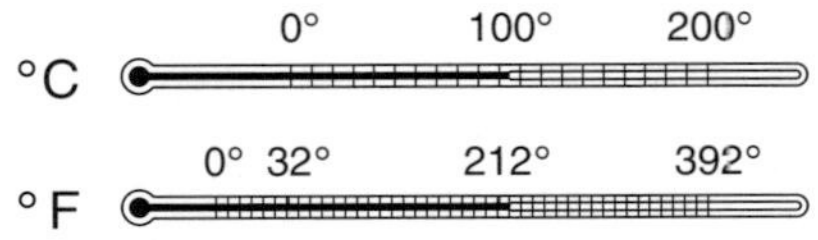

CONVERT 105°C TO °F

$$°F = (1.8 \times °C) + 32$$
$$°F = (1.8 \times 105) + 32$$
$$°F = 189 + 32$$
$$°F = 221° \ F$$

CELSIUS TO
FAHRENHEIT CONVERSION

CONVERT 221° F TO °C

$$°C = \frac{(°F - 32)}{1.8}$$
$$°C = \frac{(221 - 32)}{1.8}$$
$$°C = \frac{189}{1.8}$$
$$°C = 105° \ C$$

FAHRENHEIT TO
CELSIUS CONVERSION

Figure 7-11. Celsius and Fahrenheit are the two temperature scales most commonly used.

To convert a temperature from one scale to the other, the difference in the two bases of the scales and the ratio of the scales are considered. The base on the Celsius scale is 0°C, and the base on the Fahrenheit scale is 32°F. The 32° difference between the bases is used when converting temperatures.

The ratio between the two is determined by the difference between freezing points (0°C or 32°F) and boiling points (100°C or 212°F) of water on the two scales. There is a range of 100°C between 0°C and 100°C on the Celsius scale, and a range of 180°F between 32°F and 212°F on the Fahrenheit scale.

The ratio for the conversion is found by dividing 180 by 100, which is a 1.8 ratio. There is 1.0°C on the Celsius scale for every 1.8 °F on the Fahrenheit scale.

Celsius to Fahrenheit. To convert Celsius to Fahrenheit, multiply 1.8 by the Celsius reading and add 32. For example, to convert 105°C to Fahrenheit, multiply 1.8 by 105 (1.8 × 105 = 189). Add 32 to 189 (189 + 32 = **221°F**).

Fahrenheit to Celsius. To convert Fahrenheit to Celsius, subtract 32 from the Fahrenheit reading and divide by 1.8. For example, to convert 221°F to Celsius, subtract 32 from 221 (221 − 32 = 189). Divide 189 by 1.8 (189 ÷ 1.8 = **105°C**).

Absolute Temperature Scales. Other types of scales are absolute temperature scales. These include the Kelvin (°K) and Rankine (°R) scales. Absolute temperature scales are measured from the absolute zero temperature. *Absolute zero* is a theoretical condition where no heat is present. See Figure 7-12.

The Kelvin and Rankine scales use absolute zero as a common base. The Kelvin scale is the absolute temperature scale related to the Celsius scale. The Rankine scale is the absolute temperature scale related to the Fahrenheit scale.

Celsius to Kelvin. On the Kelvin scale, absolute zero (0°K) is 273°C below 0°C. To convert Celsius to Kelvin, add 273 to the Celsius reading. For example, to convert 20°C to Kelvin, add 273 to 20 (273 + 20 = **293°K**).

Kelvin to Celsius. To convert Kelvin to Celsius, subtract 273 from the Kelvin reading. For example, to convert 340°K to Celsius, subtract 273 from 340 (340 − 273 = **67°C**).

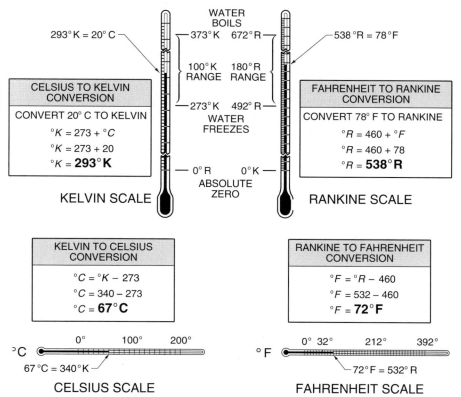

Figure 7-12. Kelvin and Rankine are two temperature scales used to measure absolute zero.

Fahrenheit to Rankine. On the Rankine scale, absolute zero (0°R) is 460°F below 0°F. To convert Fahrenheit to Rankine, add 460 to the Fahrenheit reading. For example, to convert 78°F to Rankine, add 460 to 78 (460 + 78 = **538°R**).

Rankine to Fahrenheit. To convert Rankine to Fahrenheit, subtract 460 from the Rankine reading. For example, to convert 532°R to Fahrenheit, subtract 460 from 532 (532 − 460 = **72°F**).

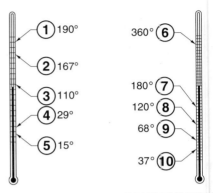

CELSIUS   FAHRENHEIT

PRACTICE PROBLEMS

Temperature

Convert °C to °F and °F to °C.

1. _____°F
2. _____°F
3. _____°F
4. _____°F
5. _____°F
6. _____°C
7. _____°C
8. _____°C
9. _____°C
10. _____°C

11. NEC® Table 310-16 permits THW Cu conductor to be installed in an ambient temperature of 75°C. Convert this figure to °F.
12. NEC Table 310-16 permits XHHW Al or Cu-clad Al to be installed in an ambient temperature of 90°C. Convert this figure to °F.
13. The recommended comfort zone for one-family dwellings is 68°F to 72°F during the daytime. Convert these figures to °C.
14. The ideal temperature when pouring concrete is 50°F to 70°F. Convert these figures to °C.
15. Convert 34°C to Kelvin.
16. Convert 300°K to Celsius.
17. Convert 96°F to Rankine.
18. Convert 495°R to Fahrenheit.

Name _____ Date _____

True-False and Completion

_____ **1.** A(n) _____ number is any number used by itself.

T   F   2. Arabic numerals are used with the SI metric system.

T   F   3. A denominate number is a concrete number that has one unit of measure.

_____ **4.** The _____ system is the most common measurement system used in the world.

_____ **5.** Cubic measure is the measurement of _____ .

_____ **6.** _____ measure is the measurement of area.

_____ **7.** _____ is the measurement of matter contained in an object.

_____ **8.** In the English system, _____ weight is the measurement of precious metals.

_____ **9.** _____ and Fahrenheit are the two temperature scales most commonly used.

_____ **10.** The Kelvin and Rankine scales use _____ as a common base.

Calculations

_____ **1.** Change 1188 sq ft to square yards.

_____ **2.** How many square feet does a ¾ A building lot contain?

_____ **3.** How many ounces are in 3 lb?

_____ **4.** Subtract 3 yd, 2′, 8″ from 12 yd, 2′, 6″.

_____ **5.** A rectangular field measures 130′ by 210′. How many square feet does the field contain?

_____ **6.** Add 6 bu, 3 pk and 4 bu, 3 pk (dry measure).

_____ **7.** A board is 6″ wide and $4\frac{1}{2}$′ long. How many square feet does the board contain?

_____ **8.** Change 512 cm to millimeters.

_____ **9.** Change 312 l to dekaliters.

_____ **10.** Convert 2 m to inches.

_____ **11.** Convert 5 l to gallons.

_____ **12.** Convert 1420 g to pounds.

_____ **13.** Convert 6″ to centimeters.

_____ **14.** Convert 3 sq mi to square kilometers.

_____ **15.** Convert 12 cu ft to cubic meters.

_____ **16.** Convert 20 m to kilometers.

_____ **17.** Convert 98°C to Fahrenheit.

_____ **18.** Convert 102°F to Celsius.

_____ **19.** Convert 310°K to Celsius.

_____ **20.** Convert 510°R to Fahrenheit.

Name _____ Date _____

_____ **1.** Change 137.5 rd to yards.

_____ **2.** Change 1½ A to square feet.

_____ **3.** Change 121.5 cu ft to cubic yards.

_____ **4.** Change 128 fl oz to quarts.

_____ **5.** Change 17 t to pounds.

_____ **6.** Multiply 3 yd, 2′, 6″ by 4.

_____ **7.** Add 3 mi, 302′, 7″ and 4 mi, 19′, 6″.

_____ **8.** Subtract 3 t, 460 lb from 16 t, 305 lb.

_____ **9.** Add 14 t, 900 lb and 27 t, 1300 lb.

_____ **10.** Convert 16 m to inches.

_____ **11.** Convert 110 km to miles.

_____ **12.** Convert 10 km^2 to square miles. *(3 places)*

_____ **13.** Convert 15 cm^2 to square inches. *(3 places)*

_____ **14.** Convert 13 cu ft to cubic yards. *(3 places)*

_____ **15.** Convert 24 cm^3 to cubic inches. *(3 places)*

181

_____ **16.** Convert 10 g to ounces.

_____ **17.** Convert 17 mi to kilometers.

_____ **18.** Convert 160 A to hectares.

_____ **19.** Convert 210 sq ft to square meters.

_____ **20.** Convert 19″ to centimeters.

_____ **21.** Convert 21 m^3 to cubic yards.

_____ **22.** Convert 38.5 cu in. to cubic centimeters.

_____ **23.** Convert 2.5 bu to liters.

_____ **24.** Convert 10.5 troy oz to grams.

_____ **25.** Convert 22°C to Kelvin.

_____ **26.** Convert 81°F to Rankine.

_____ **27.** Convert 385°K to Celsius.

_____ **28.** Convert 510°R to Fahrenheit.

_____ **29.** Convert 88°C to Fahrenheit.

_____ **30.** Convert 185°F to Celsius.

POWERS and ROOTS

A power is the product of factors that are the same number. A base is the factor that is multiplied to obtain the power. Involution is finding powers of numbers by raising the base to the indicated exponent. A root of a number is one of the equal factors of a power.

POWERS

Factors are two or more numbers multiplied together to give a product. For example, in the problem $3 \times 4 \times 6 = 72$, the 3, 4, and 6 are factors of 72. A *power* is the product of a repeated factor. A *base* is the repeated factor that is multiplied to obtain a power. For example, in the problem $3 \times 3 \times 3 \times 3 = 81$, the product (81) is the power and the factor (3) is the base. The number 3 is the base of the power 81, and 81 is the power of the base 3.

When the base is used twice as a factor, the product is to the second power or square. For example, the square of 4 is 4×4 or 16. Square is used to find the area of flat surfaces.

When the base is used three times as a factor, the product is to the third power or cube. For example, the cube of 5 is $5 \times 5 \times 5$ or 125. Cube is used to find the volume of solids. When the base is used as a factor more than three times, the product is to the fourth power, fifth power, sixth power, etc. and is indicated by an exponent. See Figure 8-1.

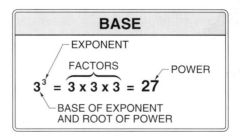

Figure 8-1. A base is the repeated factor that is multiplied to obtain the power.

An *exponent* indicates the number of times the base is raised to a power (number of times used as a

183

factor). It is indicated with a superscript. A *superscript* is a small figure above and to the right of the base. For example, 2^4 (two raised to the fourth power) indicates that base 2 is multiplied four times as a factor ($2 \times 2 \times 2 \times 2 = 16$), and 16 is the fourth power of the base 2.

Involution

Powers are a short method of writing repeated factors and indicating their product. *Involution* is the process of finding powers of numbers by raising the base to the indicated power (exponent). To find a power, a positive number, negative number, fraction, or decimal can be used as a base raised to an exponent.

An exponent may be a positive number, negative number, or zero. A power of a power is the result of a base with an exponent raised to a power. Repeated bases raised to the same or different exponents can be multiplied or divided.

Decimal and fractional exponents such as $4^{3.5}$ and $3^{2/5}$ are required in some engineering applications. Solutions to such squares, square roots, cubes, and cube roots of numbers with many digits can be worked faster with logarithms than by ordinary arithmetic.

Positive Base Numbers. To raise a positive base number to a power, multiply the positive base number by itself as many times as shown by the exponent. See Figure 8-2. For example, to solve 4^3, multiply 4 by 4 by 4 ($4 \times 4 \times 4 = 256$). The third power of 4, or 4^3, = **256**.

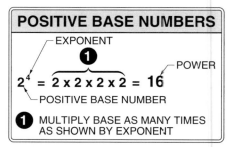

Figure 8-2. A positive base number raised to a power is multiplied as many times as shown by the exponent.

EXAMPLE

Positive Base Numbers

1. Find the value of 23^4, or the fourth power of 23.

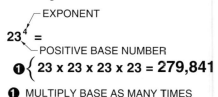

① Multiply the positive base number 23 by itself four times ($23 \times 23 \times 23 \times 23 = $ **279,841**).

Negative Base Numbers. To raise a negative number to a power, multiply the negative number by itself as many times as shown by the exponent. See Figure 8-3. If the negative number is raised to an even power, the answer is a positive number. If the negative number is raised to an odd power, the answer is a negative number.

The parentheses determine to what base the exponent applies. For example, the difference between $(-4)^2$ and $-(4)^2$ is that $(-4)^2$ is $-4 \times -4 = 16$ and $-(4)^2$ is $-(4 \times 4) = -16$.

EXAMPLES

Negative Base Numbers

1. Find the value of $(-4)^4$, or the fourth power of -4.

EVEN EXPONENT

$(-4)^4 =$ POSITIVE POWER — NEGATIVE BASE NUMBER

❶ { $-4 \times -4 \times -4 \times -4 = $ **256**

❶ MULTIPLY BASE AS MANY TIMES AS SHOWN BY EXPONENT

① Multiply the negative base number -4 by itself four times ($-4 \times -4 \times -4 \times -4 = $ **256**, a positive number).

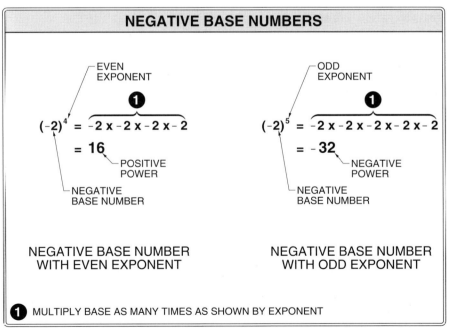

NEGATIVE BASE NUMBERS

EVEN EXPONENT

❶

$(-2)^4 = -2 \times -2 \times -2 \times -2$

$= 16$ — POSITIVE POWER

NEGATIVE BASE NUMBER

NEGATIVE BASE NUMBER WITH EVEN EXPONENT

ODD EXPONENT

❶

$(-2)^5 = -2 \times -2 \times -2 \times -2 \times -2$

$= -32$ — NEGATIVE POWER

NEGATIVE BASE NUMBER

NEGATIVE BASE NUMBER WITH ODD EXPONENT

❶ MULTIPLY BASE AS MANY TIMES AS SHOWN BY EXPONENT

Figure 8-3. A negative base number raised to a power is multiplied as many times as shown by the exponent.

2. Find the value of $(-3)^5$, or the fifth power of -3.

ODD EXPONENT

$(-3)^5 =$

NEGATIVE BASE NUMBER

❶ { $-3 \times -3 \times -3 \times -3 \times -3$

$= -729$ — NEGATIVE POWER

❶ MULTIPLY BASE AS MANY TIMES AS SHOWN BY EXPONENT

① Multiply the negative base number -3 by itself five times ($-3 \times -3 \times -3 \times -3 \times -3 = -729$, a negative number).

Fraction Base. To raise a fraction to a power, multiply the numerator and the denominator to the power separately. See Figure 8-4.

A proper fraction raised to a higher power has an answer of lesser value than the base fraction. For example, $\frac{3}{5}^2 = \frac{9}{25}$ which is less than $\frac{3}{5}$.

This is not the case with an improper fraction raised to a power. For example, $\frac{4}{3}^2 = \frac{16}{9}$ which is greater than $\frac{4}{3}$.

EXAMPLES

Fraction Base

1. Find the fifth power of $\frac{3}{4}$.

EXPONENT

❶ { $(\frac{3}{4})^5 = \frac{3^5}{4^5} = \frac{3 \times 3 \times 3 \times 3 \times 3}{4 \times 4 \times 4 \times 4 \times 4}$

PROPER FRACTION BASE $= \frac{243}{1024}$ — LESS THAN $\frac{3}{4}$

❶ MULTIPLY NUMERATOR AND DENOMINATOR SEPARATELY AS MANY TIMES AS SHOWN BY EXPONENT

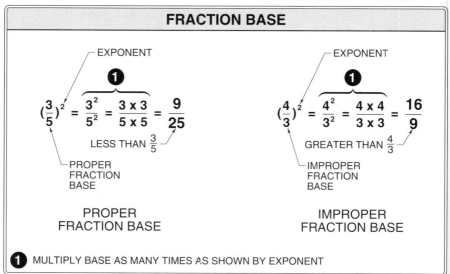

FRACTION BASE

EXPONENT

❶

$(\frac{3}{5})^2 = \frac{3^2}{5^2} = \frac{3 \times 3}{5 \times 5} = \frac{9}{25}$

LESS THAN $\frac{3}{5}$

PROPER FRACTION BASE

PROPER FRACTION BASE

EXPONENT

❶

$(\frac{4}{3})^2 = \frac{4^2}{3^2} = \frac{4 \times 4}{3 \times 3} = \frac{16}{9}$

GREATER THAN $\frac{4}{3}$

IMPROPER FRACTION BASE

IMPROPER FRACTION BASE

❶ MULTIPLY BASE AS MANY TIMES AS SHOWN BY EXPONENT

Figure 8-4. To raise a fraction to a power, multiply the numerator and denominator to the power separately.

① Raise the 3 to the fifth power and raise the 4 to the fifth power. Place the 243 over 1024 ($\frac{3 \times 3 \times 3 \times 3 \times 3}{4 \times 4 \times 4 \times 4 \times 4} = {}^{243}\!/_{1024}$).

2. Find the third power of $^5\!/_3$.

$$\mathbf{❶}\left\{ (\tfrac{5}{3})^3 = \frac{5^3}{3^3} = \frac{5 \times 5 \times 5}{3 \times 3 \times 3} \right.$$

EXPONENT

IMPROPER FRACTION BASE

$= \dfrac{\mathbf{125}}{\mathbf{27}}$ GREATER THAN $\frac{5}{3}$

❶ MULTIPLY NUMERATOR AND DENOMINATOR SEPARATELY AS MANY TIMES AS SHOWN BY EXPONENT

① Raise the 5 to the third power and raise the 3 to the third power. Place the 125 over 27 ($\frac{5 \times 5 \times 5}{3 \times 3 \times 3}$ = $^{125}\!/_{27}$).

Decimal Base. To raise decimals or mixed decimals to a power, multiply the decimal by itself as many times as indicated by the exponent. See Figure 8-5. For example, to solve $.25^3$, multiply .25 by .25 by .25 ($.25 \times .25 \times .25 = \mathbf{.015625}$).

Decimals are like proper fractions in that decimals raised to a higher power have an answer of lesser value. For example, $.2^2 = .2 \times .2 = \mathbf{.04}$ which is less than .2, and $.2^3 = .2 \times .2 \times .2 = \mathbf{.008}$ which is less in value than .2 and $.2^2$.

DECIMAL BASE

EXPONENT

$\mathbf{❶}\left\{ (.25)^3 = .25 \times .25 \times .25 \right.$

$= \mathbf{.015625}$ POWER

DECIMAL BASE

NUMBER OF DECIMAL PLACES EQUALS DECIMAL PLACES IN BASE TIMES EXPONENT

❶ MULTIPLY BASE AS MANY TIMES AS SHOWN BY EXPONENT

Figure 8-5. To raise decimals or mixed decimals to a power, multiply the decimal by itself as many times as shown by the exponent.

To find the number of decimal places in the answer, multiply the number of decimal places in the base by the exponent. For example, $.3^6 = .3 \times .3 \times .3 \times .3 \times .3 \times .3 = \mathbf{.000729}$ which has six decimal places because 1 (decimal place) multiplied by 6 (exponent) equals 6.

EXAMPLE

Decimal Base

1. Find the third power of .03.

EXPONENT

$\mathbf{❶}\left\{ (.03)^3 = .03 \times .03 \times .03 \right.$

$= \mathbf{.000027}$ POWER

DECIMAL BASE

NUMBER OF DECIMAL PLACES EQUALS DECIMAL PLACES IN BASE TIMES EXPONENT

❶ MULTIPLY BASE AS MANY TIMES AS SHOWN BY EXPONENT

① Multiply .03 by .03 by .03 ($.03 \times .03 \times .03 = \mathbf{.000027}$).

Power of a Power. The power of a power may be found by the base exponent method or the new exponent method. To use the base exponent method, solve the base with the first exponent and raise it to the second power.

To use the new exponent method, multiply the two exponents to find the new exponent which shows the number of times the base is used as a factor. See Figure 8-6.

For example, to raise 2^2 to the third power or $(2^2)^3$ using the base exponent method, solve 2^2 (base with an exponent) by multiplying 2 by 2

$(2 \times 2 = 4)$. Raise the 4 to the third power $(4^3 = 4 \times 4 \times 4 = \mathbf{64})$.

To solve the same problem by the new exponent method, multiply 2 (base exponent) by 3 (second exponent) to get 6 (new exponent).

The new problem is 2 to the sixth power or 2^6. To solve 2^6, multiply 2 by 2 by 2 by 2 by 2 by 2 $(2 \times 2 \times 2 \times 2 \times 2 \times 2 = \mathbf{64})$.

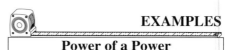

EXAMPLES

Power of a Power

1. Find the value of $(10^3)^4$ using the base exponent method.

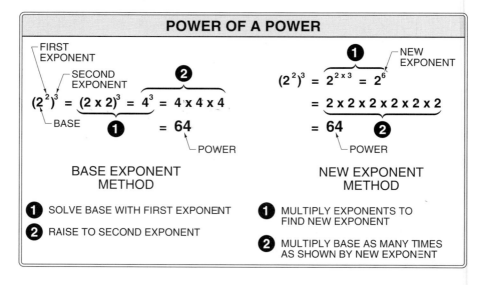

Figure 8-6. A power of a power is found by solving the base with the first exponent and raising it to the second exponent or by multiplying the exponents to find a new exponent.

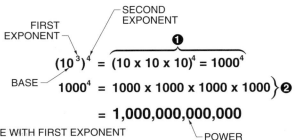

① SOLVE BASE WITH FIRST EXPONENT
② RAISE TO SECOND EXPONENT

① Solve the base with the first exponent by multiplying 10 by 10 by 10 ($10 \times 10 \times 10 = 1000$). ② Raise the 1000 to the fourth power ($1000^4 = 1000 \times 1000 \times 1000 \times 1000 = \mathbf{1{,}000{,}000{,}000{,}000}$).

2. Find the value of $(10^3)^4$ using the new exponent method.

$$(10^3)^4 = 10^{3 \times 4} = 10^{12} \text{ — NEW EXPONENT}$$

$$10^{12} = 10 \times 10 \times 10 \times 10 \times 10 \times 10 \times 10 \times 10 \times 10 \times 10 \times 10 \times 10 \Big\} ②$$

$$= 1{,}000{,}000{,}000{,}000$$

① MULTIPLY EXPONENTS TO FIND NEW EXPONENT
② MULTIPLY BASE AS MANY TIMES AS SHOWN BY NEW EXPONENT

① Multiply the exponents to find the new exponent ($10^{3 \times 4} = 10^{12}$). ② Multiply the base by 12 ($10^{12} = 10 \times 10 \times 10 \times 10 \times 10 \times 10 \times 10 \times 10 \times 10 \times 10 \times 10 \times 10 = \mathbf{1{,}000{,}000{,}000{,}000}$).

Negative Exponent. A base raised to a negative power is the reciprocal of the base with a positive exponent. See Figure 8-7.

To find the value of a base raised to a negative power, invert the base to find the reciprocal and solve the denominator. For example, to find the value of 10^{-4}, reciprocate the base and make the exponent positive ($\frac{1}{10}^4$). Solve the denominator ($\frac{1}{10}^4 = \frac{1}{10} \times \frac{1}{10} \times \frac{1}{10} \times \frac{1}{10} = \mathbf{\frac{1}{10{,}000}}$).

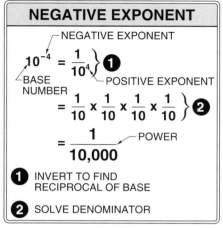

NEGATIVE EXPONENT

① INVERT TO FIND RECIPROCAL OF BASE

② SOLVE DENOMINATOR

Figure 8-7. A base raised to a negative power is the reciprocal of the base with a positive exponent.

EXAMPLE

Negative Exponent

1. Find the value of 8^{-3}.

① INVERT TO FIND RECIPROCAL OF BASE

② SOLVE DENOMINATOR

① Reciprocate the base and make the exponent positive ($8^{-3} = \frac{1}{8^3}$). ② Solve the denominator ($\frac{1}{8^3} = \frac{1}{8} \times \frac{1}{8} \times \frac{1}{8} = \frac{1}{512}$).

Zero Exponent. Any base raised to the power of zero is equal to 1. For example, $1^0 = 1$, $14^0 = 1$, and $829^0 = 1$.

Multiplication of Powers. To multiply two or more like bases with exponents, add the exponents to find the new exponent for the same base. See Figure 8-8. For example, to multiply 5^2 and 5^3, add the 2 and 3 for the new exponent ($2 + 3 = 5$). Place the 5 as the exponent next to the base and solve ($5^2 \times 5^3 = 5^{2+3} = 5^5 = \textbf{3125}$).

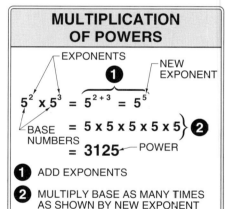

MULTIPLICATION OF POWERS

① ADD EXPONENTS

② MULTIPLY BASE AS MANY TIMES AS SHOWN BY NEW EXPONENT

Figure 8-8. To multiply two or more like bases with exponents, add the exponents to find a new exponent for the same base.

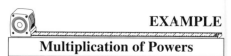

EXAMPLE

Multiplication of Powers

1. $4^2 \times 4^2 \times 4^3$.

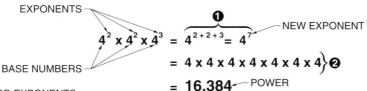

EXPONENTS

❶ NEW EXPONENT

$$4^2 \times 4^2 \times 4^3 = 4^{2+2+3} = 4^7$$

BASE NUMBERS

$$= 4 \times 4 \times 4 \times 4 \times 4 \times 4 \times 4 \} ❷$$

$$= 16,384 \leftarrow \text{POWER}$$

❶ ADD EXPONENTS

❷ MULTIPLY BASE AS MANY TIMES AS SHOWN BY NEW EXPONENT

① Add exponents ($2 + 2 + 3 = 7$). ② Multiply the base seven times ($4^7 = 4 \times 4 \times 4 \times 4 \times 4 \times 4 \times 4 = \mathbf{16,384}$).

Division of Powers. To divide two like bases with exponents, subtract the exponent of the divisor from the exponent of the dividend to find the new exponent for the same base. See Figure 8-9.

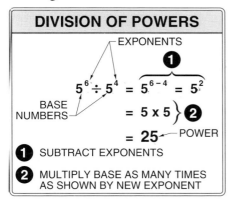

DIVISION OF POWERS

EXPONENTS

❶

$$5^6 \div 5^4 = 5^{6-4} = 5^2$$

BASE NUMBERS

$$= 5 \times 5 \} ❷$$

$$= 25 \leftarrow \text{POWER}$$

❶ SUBTRACT EXPONENTS

❷ MULTIPLY BASE AS MANY TIMES AS SHOWN BY NEW EXPONENT

Figure 8-9. To divide two like bases with exponents, subtract the exponent of the divisor from the exponent of the dividend to find the new exponent for the same base.

For example, to divide 5^6 by 5^4, subtract 4 from 6 to find the new exponent ($6 - 4 = 2$). Place the 2 as the exponent next to the base and solve ($5^6 \div 5^4 = 5^{6-4} = 5^2 = 5 \times 5 = \mathbf{25}$).

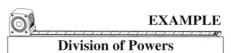

EXAMPLE

Division of Powers

1. $3^9 \div 3^5$.

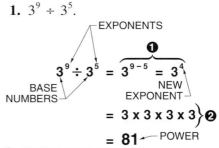

EXPONENTS

❶

$$3^9 \div 3^5 = 3^{9-5} = 3^4$$

BASE NUMBERS

NEW EXPONENT

$$= 3 \times 3 \times 3 \times 3 \} ❷$$

$$= 81 \leftarrow \text{POWER}$$

❶ SUBTRACT EXPONENTS

❷ MULTIPLY BASE AS MANY TIMES AS SHOWN BY NEW EXPONENT

① Subtract exponents ($9 - 5 = 4$). ② Place the 4 as the exponent next to the base and solve ($3^9 \div 3^5 = 3^{9-5} = 3^4 = 3 \times 3 \times 3 \times 3 = \mathbf{81}$).

PRACTICE PROBLEMS

Involution

1. Find the value of 2^6.

2. Find the fourth power of 21.

3. Raise $\frac{2}{9}$ to the third power.

4. Find the value of $(2^4)^3$.

5. Find the value of $(3^3)^{-2}$.

6. Find the volume (in cubic inches) of the Carton which is a cubic container. *Note:* Volume (V) of a cubic container equals one side raised to the third power $(V = s^3)$.

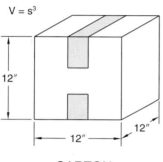

CARTON

7. Find the value of 144^0.

8. Find the surface area (in square feet) of the Supply Room. *Note:* Area (A) of a square equals one side raised to the second power $(A = s^2)$.

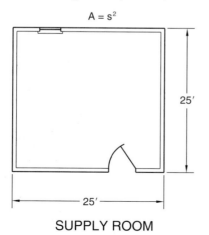

SUPPLY ROOM

9. Find the value of $(-2)^{-2}$.

10. Find the value of $(12^0)^{-3}$.

11. Find the value of $3^3 \times 3^4$.

12. Find the value of $.2^3 \times .2^3$. *(6 places)*

13. Find the value of $2.2^5 \div 2.2^2$. *(3 places)*

14. Find the value of $750^8 \div 750^6$.

15. Find the value of $(\frac{2}{5})^2 \times (\frac{2}{5})^3$.

ROOTS

The *radical sign* $(\sqrt{\ })$ is the symbol used to indicate the root of a number. A *root* of a number is one of the equal factors of a power. When the roots of a number are multiplied together, the number results. Therefore, one of the equal factors used to obtain a power is a root of the power.

For example, the two equal factors of 16 are 4 and 4 because 16 = 4 × 4. Therefore, 4 is a root of 16. Similarly, the two equal factors of 81 are 9 and 9 (81 = 9 × 9). Therefore, 9 is a root of 81.

The *radicand* is the number under the radical sign from which the root is to be found ($\sqrt{16}$). The *index* is the small figure to the left of the root sign ($\sqrt[x]{\ }$).

A *square root* is two equal factors or roots of a number. When the radical sign is used without an index it is understood to mean square root.

Thus, $\sqrt{16}$ means the square root of 16; $\sqrt{144}$ means the square root of 144. See Figure 8-10.

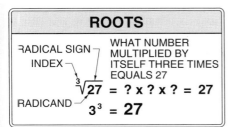

ROOTS

RADICAL SIGN ⌐   WHAT NUMBER
INDEX ⌐   MULTIPLIED BY
         ITSELF THREE TIMES
         EQUALS 27

$$\sqrt[3]{27} = ? \times ? \times ? = 27$$

RADICAND ⌐ $3^3 = 27$

Figure 8-10. When the roots of a number are multiplied, the number results.

The *cube root* is a number which has three equal roots or factors. For example, $27 = 3 \times 3 \times 3$, so 3 is the cube root of 27. When the cube root is to be found, a superscript figure 3 is placed by the radical sign indicating that $\sqrt[3]{27}$ is the cube root of 27.

When the fourth root is to be found, a superscript figure 4 is similarly placed to indicate that $\sqrt[4]{81}$ is the root of 81. If higher roots are used, the corresponding figure is used accordingly. For example, the fifth root ($\sqrt[5]{x}$), the sixth root ($\sqrt[6]{x}$), etc.

Note: If the area of a square surface is known, the length of one side is found by extracting the square root of the area. If the volume of a cube is known, the length of one side is found by extracting the cube root of the volume.

Evolution

Evolution is the general process of finding roots of numbers. *Extracting* the square root is the process of finding one of the two equal factors of a number. *Perfect squares* are numbers that have whole number square roots. Most numbers do not have whole number square roots.

Square roots of numbers that are perfect squares and have only one, two, and sometimes three figures can easily be found without any special process. In the case of large numbers and non-perfect squares, a special evolution process is necessary.

Square Root of a Whole Number. To extract the square root of a whole number, separate the number into periods of two figures. See Figure 8-11.

A zero may be annexed to complete a period. The number of periods formed is the number of figures in the square root.

Starting from the left, find the largest perfect square less than or equal to each period. Write the first square above the first period and subtract the product of the square from the period.

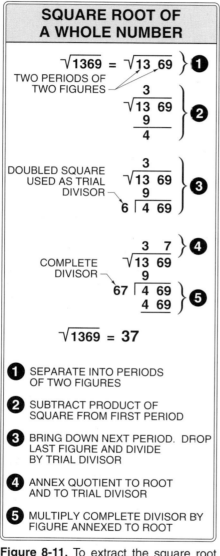

SQUARE ROOT OF A WHOLE NUMBER

$\sqrt{1369} = \sqrt{13\ 69}$ } **1**

TWO PERIODS OF TWO FIGURES

3
$\sqrt{13\ 69}$
9
4
} **2**

DOUBLED SQUARE USED AS TRIAL DIVISOR

3
$\sqrt{13\ 69}$
9
6 | 4 69
} **3**

COMPLETE DIVISOR

3 7
$\sqrt{13\ 69}$
9
67 | 4 69
4 69
} **4** **5**

$\sqrt{1369} = 37$

1 SEPARATE INTO PERIODS OF TWO FIGURES

2 SUBTRACT PRODUCT OF SQUARE FROM FIRST PERIOD

3 BRING DOWN NEXT PERIOD. DROP LAST FIGURE AND DIVIDE BY TRIAL DIVISOR

4 ANNEX QUOTIENT TO ROOT AND TO TRIAL DIVISOR

5 MULTIPLY COMPLETE DIVISOR BY FIGURE ANNEXED TO ROOT

Figure 8-11. To extract the square root of a whole number, separate the number into periods of two figures.

Bring down the next period and place it by the remainder. Double the square and use it as a trial divisor.

Place the trial divisor to the left of the remainder. Disregard the last figure of the remainder and divide by the trial divisor. Annex the quotient as the next figure in the square root and annex it to the trial divisor.

Multiply the complete divisor by the figure annexed to the root. Place the product under the remainder and subtract. Repeat until no remainder is found or the desired number of decimal places is found.

For example, to find the square root of 1369, separate the $\sqrt{1369}$ into periods of two figures ($\sqrt{13\ 69}$). The number of periods formed (2) is the same as the number of figures in the square root.

Find the largest perfect square less than or equal to the first period. The largest perfect square of 13 is 3. Write the 3 above the 13 and subtract 9 (product of the square) from the 13 to get 4.

Bring down the second period and place it by 4 to get 469. Double the square (3) to get 6 and use it as a trial divisor. Place the 6 to the left of the remainder. Disregard the last digit of the remainder to get 46, and divide by the trial divisor to get 7.

Annex the 7 as the next figure in the square root and the trial divisor (67). Multiply the 67 by 7 (67

× 7 = 469). Subtract 469 from 469. There is no remainder. The square root of 1369 = **37**.

EXAMPLE

| Square Root of a Whole Number |

1. Find the square root of 441.

ANNEXED
ZERO

$$\sqrt{441} = \sqrt{04\,41} \Big\} \text{❶}$$

TWO PERIODS OF
TWO FIGURES

$$
\begin{array}{c}
2 \\
\sqrt{04\ 01} \\
4 \\
\hline
0
\end{array}
\Big\} \text{❷}
$$

DOUBLED SQUARE
USED AS TRIAL
DIVISOR

$$
\begin{array}{c}
2 \\
\sqrt{04\ 41} \\
4 \\
\hline
4\ \lceil 0\ 41
\end{array}
\Big\} \text{❸}
$$

COMPLETE
DIVISOR

$$
\begin{array}{c}
2\quad 1 \\
\sqrt{04\ 41} \\
4 \\
\hline
41\ \lceil 0\ 41 \\
41 \\
\hline
0
\end{array}
\Big\} \text{❹} \Big\} \text{❺}
$$

$$\sqrt{441} = 21$$

❶ SEPARATE INTO PERIODS
OF TWO FIGURES

❷ SUBTRACT PRODUCT OF
SQUARE FROM FIRST PERIOD

❸ BRING DOWN NEXT PERIOD. DROP
LAST FIGURE AND DIVIDE
BY TRIAL DIVISOR

❹ ANNEX QUOTIENT TO ROOT
AND TO TRIAL DIVISOR

❺ MULTIPLY COMPLETE DIVISOR BY
FIGURE ANNEXED TO ROOT

① Annex a zero and separate the $\sqrt{441}$ into periods of two figures ($\sqrt{04\,41}$). The number of periods formed (2) is the same as the number of figures in the square root.

② Find the largest perfect square less than or equal to 4. The largest perfect square is 2. Write the 2 above the first period and subtract 4 (product of the square) from the 4 to get 0.

③ Bring down the 41 and place it by the 0 (041). Double the square (2) to get 4 and use it as a trial divisor. Place the 4 to the left of the remainder. Disregard the last digit (1) of the remainder to get 04, and divide by the trial divisor to get 1.

④ Annex the 1 as the next figure in the square root and the trial divisor (41).

⑤ Multiply the 41 by 1 (41 × 1 = 41). Subtract 41 from 41 (41 − 41 = 0). There is no remainder. The square root of 441 = **21**.

Square Root of a Decimal Number. To find the square root of a decimal number, place a new decimal point above the old decimal point. Separate the number into periods of two figures to the left and the right of the decimal. See Figure 8-12.

The two figures of a period must never be separated by the decimal point. A zero may be annexed to complete a period.

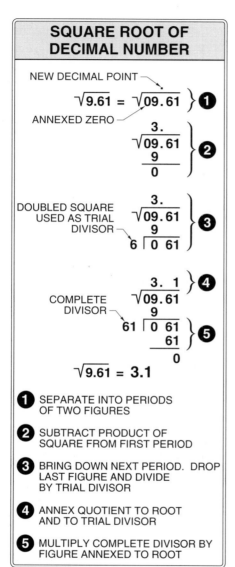

SQUARE ROOT OF DECIMAL NUMBER

NEW DECIMAL POINT

$\sqrt{9.61} = \sqrt{09.61}$ **①**

ANNEXED ZERO

3.
$\sqrt{09.61}$ **②**
9
0

DOUBLED SQUARE USED AS TRIAL DIVISOR

3.
$\sqrt{09.61}$ **③**
9
6⌐0 61

COMPLETE DIVISOR

3. 1 **④**
$\sqrt{09.61}$
9
61⌐0 61
61 **⑤**
0

$\sqrt{9.61} = 3.1$

① SEPARATE INTO PERIODS OF TWO FIGURES

② SUBTRACT PRODUCT OF SQUARE FROM FIRST PERIOD

③ BRING DOWN NEXT PERIOD. DROP LAST FIGURE AND DIVIDE BY TRIAL DIVISOR

④ ANNEX QUOTIENT TO ROOT AND TO TRIAL DIVISOR

⑤ MULTIPLY COMPLETE DIVISOR BY FIGURE ANNEXED TO ROOT

Figure 8-12. To find the square root of a decimal number, place a new decimal point above the old decimal point and separate the number into periods of two figures.

After placing the new decimal point, the remaining steps of find-ing the square root of a decimal number are the same as finding the square root of a whole number.

For example, to find the square root of 9.61, annex a zero, place a new decimal point above the old decimal point, and separate the $\sqrt{09.61}$ into periods of two figures ($\sqrt{09.61}$). The number of periods formed (2) is the same as the num-ber of figures in the square root.

Find the largest perfect square less than or equal to 9. The largest perfect square is 3. Write the 3 above the first period and subtract 9 (product of the square) from the 9 to get 0.

Bring down the 61 and place it by 0 (0 61). Double the square (3) to get 6 and use it as a trial divisor. Place the 6 to the left of the re-mainder. Disregard the last digit of the remainder to get 6 and divide by the trial divisor to get 1.

Annex the 1 as the next figure in the square root and the trial di-visor (61). Multiply the 61 by 1 (61 × 1 = 61). Subtract 61 from 61. There is no remainder. The square root of 9.61 = **3.1**.

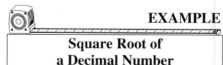

EXAMPLE

Square Root of a Decimal Number

1. Find the square root of .001225.

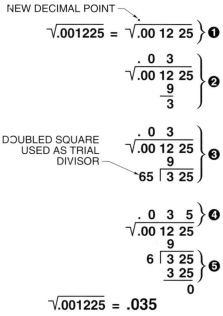

NEW DECIMAL POINT

$\sqrt{.001225}$ = $\sqrt{.00\ 12\ 25}$ } **❶**

```
         . 0  3
  √.00 12 25        } ❷
          9
          3
```

DOUBLED SQUARE
USED AS TRIAL
DIVISOR

```
          . 0  3
  √.00 12 25        } ❸
           9
   65 ⌐3 25
```

```
          . 0  3  5   } ❹
  √.00 12 25
           9
   6 ⌐3 25            } ❺
     3 25
          0
```

$\sqrt{.001225}$ = **.035**

❶ SEPARATE INTO PERIODS OF TWO FIGURES

❷ SUBTRACT PRODUCT OF SQUARE FROM FIRST PERIOD

❸ BRING DOWN NEXT PERIOD. DROP LAST FIGURE AND DIVIDE BY TRIAL DIVISOR

❹ ANNEX QUOTIENT TO ROOT AND TO TRIAL DIVISOR

❺ MULTIPLY COMPLETE DIVISOR BY FIGURE ANNEXED TO ROOT

① Separate the $\sqrt{.001225}$ into periods of two figures ($\sqrt{.00\ 12\ 25}$) and place a new decimal above the cld. The number of periods formed (3) is the same as the number of figures in the square root. The first period contains only zeros, so 0 is the first figure in the square root.

② Find the largest perfect square less than or equal to the next period

12. The largest perfect square is 3. Write the 3 as the next figure in the square root and subtract 9 (product of the square) from the 12 (12 − 9 = 3).

③ Bring down the 25 and place it by the 3 (325). Double the square 3 to get 6 and use it as a trial divisor. Place the 6 to the left of the remainder. Disregard the last digit of the remainder (32) and divide by the trial divisor (32 ÷ 6 = 5).

④ Annex the 5 as the next figure in the square root and to the trial divisor (65).

⑤ Multiply the 65 by 5 (65 × 5 = 325). Subtract 325 from 325. There is no remainder. The square root of .001225 = **.035**.

Square Root of a Fraction. A fraction may or may not have numerators and denominators that are perfect squares. See Figure 8-13.

To find the square root of a fraction when the numerator and denominator are perfect squares, find the square root of the numerator and the denominator separately. For example, to find the square root of $\frac{4}{25}$, find the square root of 4 and find the square root of 25 ($\sqrt{\frac{4}{25}}$ = $\frac{2}{5}$).

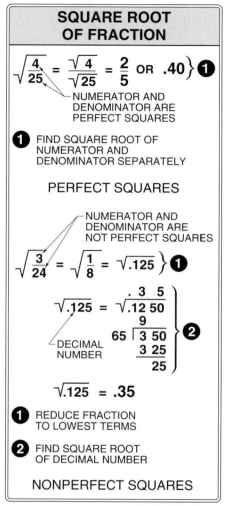

SQUARE ROOT OF FRACTION

$$\sqrt{\frac{4}{25}} = \frac{\sqrt{4}}{\sqrt{25}} = \frac{2}{5} \text{ OR } .40 \Big\} \text{①}$$

NUMERATOR AND DENOMINATOR ARE PERFECT SQUARES

① FIND SQUARE ROOT OF NUMERATOR AND DENOMINATOR SEPARATELY

PERFECT SQUARES

NUMERATOR AND DENOMINATOR ARE NOT PERFECT SQUARES

$$\sqrt{\frac{3}{24}} = \sqrt{\frac{1}{8}} = \sqrt{.125} \Big\} \text{①}$$

$$\sqrt{.125} = \sqrt{.12\ 50}$$

DECIMAL NUMBER

$$
\begin{array}{r}
.\ 3\ 5 \\
\sqrt{.12\ 50} \\
9 \\
65\ \overline{|\ 3\ 50} \\
3\ 25 \\
25
\end{array} \Big\} \text{②}
$$

$$\sqrt{.125} = .35$$

① REDUCE FRACTION TO LOWEST TERMS

② FIND SQUARE ROOT OF DECIMAL NUMBER

NONPERFECT SQUARES

Figure 8-13. To find the square root of a fraction, first determine if the fraction has a numerator and denominator that are perfect squares.

To find the square root of a fraction when the numerator and denominator are not perfect squares, reduce the fraction to lowest terms, change reduced fraction to a decimal number (divide the numerator by the denominator), and find the square root of the decimal number. When finding the square root of a decimal number, repeat the process until there is no remainder or the desired number of decimal places is found.

For example, to find the square root of $\frac{3}{24}$ to the second decimal place, reduce the fraction by dividing 3 and 24 by 3 ($\frac{1}{8}$). Divide 1 (numerator) by 8 (denominator) (1 ÷ 8 = .125).

To find the square root of .125, separate the $\sqrt{.125}$ into periods of two figures ($\sqrt{.12\ 50}$) and place a new decimal above the old decimal. A zero is annexed to the 5 to make a two figure period.

Find the largest perfect square less than or equal to 12. The largest perfect square is 3. Write the 3 above the first period and subtract 9 (product of the square) from the 12 (12 − 9 = 3).

Bring down the 50 and place it by 3 (350). Double the square (3) to get 6 and use it as a trial divisor. Place the 6 to the left of the remainder. Disregard the last digit of the remainder (35) and divide by the trial divisor (35 ÷ 6 = 5). Annex the 5 as the next figure in the square root and to the trial divisor (65).

Multiply the 65 by 6 (65 × 6 = 325). Subtract 325 from 350 (350 −

325 = 25). Since two decimal places are found, **.35** is the square root of $\frac{7}{24}$ to the second decimal place.

EXAMPLES

Square Root of a Fraction

1. Find the square root of $\frac{9}{16}$.

$$\sqrt{\frac{9}{16}} = \frac{\sqrt{9}}{\sqrt{16}} = \frac{3}{4} \text{ OR } .75 \Big\} \text{❶}$$

NUMERATOR AND
DENOMINATOR ARE
PERFECT SQUARES

❶ FIND SQUARE ROOT OF
NUMERATOR AND
DENOMINATOR SEPARATELY

The numerator and denominator are perfect squares. ① Find the square root of the 9 and the 16 $\sqrt{\frac{9}{16}} = \sqrt{9} \sqrt{16} = \frac{3}{4}$ or .75).

2. Find the square root of $\frac{3}{4}$ to the second decimal place.

$$\sqrt{\frac{3}{4}} = \sqrt{.75} \Big\} \text{❶}$$

$$
\begin{array}{r}
.\ 8\ \ 6 \\
\sqrt{.75\ 00} \\
64 \\
166\ \overline{|11\ 00} \\
9\ 96 \\
1\ 04
\end{array}
\Bigg\} \text{❷}
$$

$\sqrt{.75} = .86$

❶ REDUCE FRACTION TO LOWEST TERMS
❷ FIND SQUARE ROOT
OF DECIMAL NUMBER

① The fraction is reduced to lowest terms. Divide the 3 by the 4 (3 ÷ 4 = .75). To find the square root of .75, separate the $\sqrt{.75}$ into periods of two figures ($\sqrt{.75\ 00}$) and place a new decimal above the old.

② Find the largest perfect square less than or equal to 75. The largest perfect square is 8. Write the 8 above the first period and subtract 64 (product of the square) from the 75 (75 − 64 = 11). Bring down two zeros and place them by 11 (1100). Double the square (8) to get 16 and use it as a trial divisor. Place the 16 to the left of the remainder. Disregard the last digit of the remainder (110) and divide by the trial divisor (110 ÷ 16 = 6).

Annex the 6 as the next figure in the square root and the trial divisor (166). Multiply the 166 by 6 (166 × 6 = 996). Subtract 996 from 1100 (1100 − 996 = 104). Since two decimal places are found, **.86** is the square root of $\frac{3}{4}$ to the second decimal place.

Powers and Roots Table. A second method of evolution is through the use of powers and roots tables. See Figure 8-14. See Appendix. Most powers and roots tables not only list the square roots of numbers, but also list squares, cubes, cube roots, and reciprocals of numbers.

POWERS AND ROOTS TABLE						
No.	Square	Cube	Sq. Root	Cube Root	Reciprocal	No.
1	1	1	1.00000	1.00000	1.0000000	1
2	4	8	1.41421	1.25992	.5000000	2
3	9	27	1.73205	1.44225	.3333333	3
4	16	64	2.00000	1.58740	.2500000	4
5	25	125	2.23607	1.70998	.2000000	5
6	36	216	2.44949	1.81712	.1666667	6
7	49	343	2.64575	1.91293	.1428571	7
8	64	512	2.82843	2.00000	.1250000	8
9	81	729	3.00000	2.08008	.1111111	9
10	100	1000	3.16228	2.15443	.1000000	10
11	121	1331	3.31662	2.22398	.0909091	11
12	144	1728	3.46410	2.28943	.0833333	12
13	169	2197	3.60555	2.35133	.0769231	13
14	196	2744	3.74166	2.41014	.0714286	14
15	225	3375	3.87298	2.46621	.0666667	15
16	256	4096	4.00000	2.51984	.0625000	16
17	289	4913	4.12311	2.57128	.0588235	17
18	324	5832	4.24264	2.62074	.0555556	18
19	361	6859	4.35890	2.66840	.0526316	19
20	400	8000	4.47214	2.71442	.0500000	20
21	441	9261	4.58258	2.75892	.0476190	21
22	484	10,648	4.69042	2.80204	.0454545	22
23	529	12,167	4.79583	2.84387	.0434783	23
24	576	13,824	4.89898	2.88450	.0416667	24
25	625	15,625	5.00000	2.92402	.0400000	25

Figure 8-14. The Powers and Roots Table is used to find square roots of whole numbers.

To find the square, cube, etc., find the number in the *No.* column and move horizontally to the appropriate column. For example, to find the square root of 11, find 11 in the *No.* column and move horizontally to the *Sq. Root* column. The square root of 11 is **3.31662**.

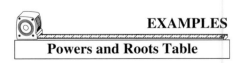

EXAMPLES

Powers and Roots Table

1. Using the Powers and Roots Table, find the square root of 17.

POWERS AND ROOTS TABLE						
No.	Square	Cube	Sq. Root	Cube Root	Reciprocal	No.
14	196	2744	3.74166	2.41014	.0714286	14
15	225	3375	3.87298	2.46621	.0666667	15
16	256	4096	4.00000	2.51984	.0625000	16
17	289	4913	4.12311	2.57128	.0588235	17
18	324	5832	4.24264	2.62074	.0555556	18
19	361	6859	4.35890	2.66840	.0526316	19

$$\sqrt{17} = 4.12311$$

❶ FIND NUMBERS AND MOVE TO APPROPRIATE COLUMN

① Find 17 in the *No.* column and move horizontally to the *Sq. Root* column. The square root of 17 is **4.12311**.

2. Using the Powers and Roots Table, find the square of 25.

POWERS AND ROOTS TABLE						
No.	Square	Cube	Sq. Root	Cube Root	Reciprocal	No.
20	400	8000	4.47214	2.71442	.0500000	20
21	441	9261	4.58258	2.75892	.0476190	21
22	484	10,648	4.69042	2.80204	.0454545	22
23	529	12,167	4.79583	2.84387	.0434783	23
24	576	13,824	4.89898	2.88450	.0416667	24
25	625	15,625	5.00000	2.92402	.0400000	25

❶ FIND NUMBERS AND MOVE TO APPROPRIATE COLUMN

② Find 25 in the *No.* column and move horizontally to the *Square* column. The square of 25 is **625**.

Fractional Exponents. A number may have an exponent that is a fraction such as $2^{4/2}$. The numerator of the fractional exponent indicates the power while the denominator of the exponent indicates the root index. See Figure 8-15. For example, to find the value of $2^{4/2}$, find the square root for the fourth power of 2 ($\sqrt{2^4}$, or $\sqrt{16}$). The square root of 16 is 4 ($2^{4/2} = \sqrt{2^4} = \sqrt{16} = 4$).

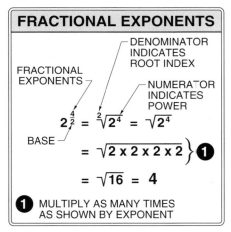

FRACTIONAL EXPONENTS

Figure 8-15. The numerator of the fractional exponent indicates the power while the denominator indicates the root index.

EXAMPLE

Fractional Exponents

1. Find the value of $4^{5/2}$.

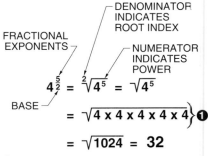

$$4^{\frac{5}{2}} = \sqrt[2]{4^5} = \sqrt{4^5}$$
$$= \sqrt{4 \times 4 \times 4 \times 4 \times 4} \}\ ❶$$
$$= \sqrt{1024} = 32$$

❶ MULTIPLY BASE AS MANY TIMES AS SHOWN BY EXPONENT

① Find the square root for the fifth power of 4 ($\sqrt{4^5}$ or $\sqrt{1024}$). The square root of 1024 is 32 ($4^{5/2} = \sqrt{4^5} = \sqrt{1024} = 32$).

PRACTICE PROBLEMS

Evolution

1. Find the square root of 1444.
2. Find the square root of 7.5625.
3. Find the square root of .1849.
4. Find the square root of 225.
5. Find the square root of .0361.
6. Find the square root of 1936.
7. Find the square root of 85. *(5 places)*
8. Find the square root of 22.5625.
9. Find the square root of 961.
10. Find the square root of 2304.
11. Find the square root of 650.25.
12. Find the square root of 15,876.
13. Find the square root of 3.0625.
14. Find the square root of 2401.

Use the Powers and Roots Table for problems 15 through 20.

15. Find the square root of 32. *(5 places)*

16. Find the square root of 39. *(3 places)*

17. Find the square root of 37. *(5 places)*

18. Find the reciprocal of 33. *(5 places)*

19. Find the cube of 40.

20. Find the cube root of 40. *(5 places)*

POWERS AND ROOTS TABLE

No.	Square	Cube	Sq Root	Cu Root	Reciprocal	No.
31	961	29,791	5.56776	3.14138	.0322581	31
32	1024	32,768	5.65685	3.17480	.0312500	32
33	1089	35,937	5.74456	3.20753	.0303030	33
34	1156	39,304	5.83095	3.23961	.0294118	34
35	1225	42,875	5.91608	3.27107	.0285714	35
36	1296	46,656	6.00000	3.30193	.0277778	36
37	1369	50,653	6.08276	3.33222	.0270270	37
38	1444	54,872	6.16441	3.36198	.0263158	38
39	1521	59,319	6.24500	3.39121	.0256410	39
40	1600	64,000	6.32456	3.41995	.0250000	40

Name _____ Date _____

True-False and Completion

_____ **1.** A(n) _____ is the product of a repeated factor.

_____ **2.** _____ is the process of finding powers of numbers by raising the base to the indicated powers.

T   F   3. An exponent is indicated with a superscript.

T   F   4. Any base raised to the power of zero is equal to zero.

T   F   5. Square is used to find the area of flat surfaces when the base is used twice as a factor.

_____ **6.** _____ is the process of finding roots of numbers.

_____ **7.** A(n) _____ is a small number above and to the right of the base number.

_____ **8.** A(n) _____ indicates the number of times that the base is raised to a power.

_____ **9.** The _____ is the small figure to the left of the root sign.

_____ **10.** _____ are two or more numbers multiplied together to give a product.

T   F   11. An exponent may be a positive number, negative number, or zero.

T   F   12. If a negative number is raised to an odd power, the answer is a positive number.

T   F   13. A proper fraction raised to a higher power has an answer of lesser value.

T   F   14. A root of a number is one of the equal factors of a power.

T   F   15. Perfect squares always have whole number square roots.

205

Calculations

_____ **1.** Find the value of 14^4.

_____ **2.** Find the value of $(-5)^4$.

_____ **3.** Find the value of $-(5)^4$.

_____ **4.** Find the fourth power of $\frac{5}{8}$.

_____ **5.** Find the third power of $\frac{3}{16}$.

_____ **6.** Find the third power of .15. *(6 places)*

_____ **7.** Find the value of $(2^4)^3$ using the base exponent method.

_____ **8.** Find the value of $(10^3)^2$ using the new exponent method.

_____ **9.** Find the value of 9^{-2}.

_____ **10.** Multiply 5^3 by 5^3 by 5^5.

_____ **11.** Multiply 4^2 by 4^3 by 4^3.

_____ **12.** Divide 4^7 by 4^3.

_____ **13.** Divide 3^5 by 3^2.

_____ **14.** Find the value of $.3^3 \times .3^3$. *(6 places)*

_____ **15.** Find the square root of 1156.

_____ **16.** Find the square root of 1764.

_____ **17.** Find the square root of 10.89.

_____ **18.** Find the square root of $\frac{4}{25}$.

Use the Powers and Roots Tables in the Appendix for the following.

_____ **19.** Find the square root of 24.

_____ **20.** Find the cube root of 17.

Name _____ Date _____

_____ **1.** Find the value of 14^4.

_____ **2.** Find the value of 7^4.

_____ **3.** Find the value of $(-5)^3$.

_____ **4.** Find the fourth power of -5.

_____ **5.** Find the third power of $\frac{1}{2}$.

_____ **6.** Find the third power of $\frac{2}{3}$.

_____ **7.** Change $\frac{3}{4}$ to a decimal and find the third power. *(6 places)*

_____ **8.** Change .25 to a fraction and find the third power.

_____ **9.** Find the value of $(6^3)^2$ using the base exponent method.

_____ **10.** Find the value of $(7^3)^2$ using the new exponent method.

_____ **11.** Find the value of $(3^2)^2$.

_____ **12.** Find the value of $(.5^3)^2$. *(6 places)*

_____ **13.** Find the value of 13×3^3.

_____ **14.** Multiply 4^3 by 4^2 by 4^2.

_____ **15.** Multiply 3^3 by 3^3 by 3^4.

_____ **16.** Multiply the fourth power of 4 times the square of 4.

_____ **17.** Divide the fifth power of 4 by the square of 4.

_____ **18.** Divide the third power of 12 by the cube of 12.

_____ **19.** Divide 3^7 by 3^5.

_____ **20.** Find the value of $(2^3)^4$.

_____ **21.** How many cubic feet does a carton with each side measuring 18″ contain? _(3 places)_

_____ **22.** How many square feet does a building lot with dimensions of 132′– 6″ on each side contain?

_____ **23.** Find the square root of 1024.

_____ **24.** Find the square root of 22. _(5 places)_

_____ **25.** Find the square root of 1369.

Use the Powers and Roots Table in the Appendix for the following.

_____ **26.** Find the cube root of 20. _(5 places)_

_____ **27.** Add $18^2 + 30^2 + 15^2$.

_____ **28.** Add $12^3 + 21^2$.

_____ **29.** Find the reciprocal of 19. _(5 places)_

_____ **30.** Subtract $9^3 - 6^3$.

RATIOS and PROPORTIONS

A ratio is the relationship between two quantities or terms. Ratios may be simple, direct, inverse, or compound and contain fractions, decimals, and/or whole numbers. All proportions are composed of two equal ratio expressions.

RATIO

Ratio is the relationship between two quantities or terms. It is the mathematical method of making a comparison. *Relation* is how much larger or smaller one term is in comparison with another term.

Any two terms being compared must be of the same kind. Picture frames and vacuum cleaners or bolts and money cannot be compared because they are not alike. Terms must be changed into common units of measure before a comparison can be made. If they cannot be reduced to a common unit, no comparison is possible.

In all ratio calculations, two terms are known. The colon (:) is the symbol used to indicate a relation between terms. See Figure 9-1. For example, the ratio of 8 to 2 is written as the ratio expression 8 : 2.

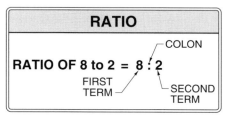

Figure 9-1. The colon is the symbol used to indicate a relation between terms.

To find the ratio of two terms, divide the first term by the second term. A ratio can be expressed as a fraction where the first term is the numerator and the second term is the denominator. For example, the ratio of 15 to 5 = 15 : 5 = 15 ÷ 5 = $\frac{15}{5}$ = **3**.

Simple Ratios

A simple ratio is the ratio between two terms. A simple ratio where the first term is smaller than the second term is similar to a proper fraction.

The answer is less than 1, resulting in a decimal hundredth or a fraction. For example, in 5 : 10 the first term (5) is smaller than the second term (10) and equals 5 divided by 10, or ½.

A simple ratio where the first term is larger than the second term is similar to an improper fraction. The answer is either a whole number or mixed number. For example, in 10 : 5 the first term (10) is larger than the second term (5) and equals 10 divided by 5, or 2.

The two types of simple ratios are direct ratios and inverse ratios. Two simple ratios can be combined to form a compound ratio. Simple ratios become more complex when one of the terms is a fraction.

Direct Ratio. A *direct ratio* is a simple ratio that results when the first term is divided by the second term. See Figure 9-2.

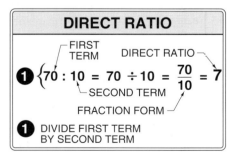

DIRECT RATIO

❶ $\{70 : 10 = 70 \div 10 = \dfrac{70}{10} = 7$

FIRST TERM, SECOND TERM, DIRECT RATIO, FRACTION FORM

❶ DIVIDE FIRST TERM BY SECOND TERM

Figure 9-2. A direct ratio is the simple ratio that results when the first term is divided by the second term.

For example, to find the direct ratio of 70 : 10, divide the first term, 70, by the second term, 10, $(70 \div 10 = 7)$.

Note: When a specific ratio expression is not named, a direct ratio should be found. For example, "find the ratio of 2 to 1" means "find the *direct* ratio of 2 to 1."

EXAMPLE

Direct Ratio

1. Find the ratio of 21 to 3.

❶ $\{21 : 3 = 21 \div 3 = \dfrac{21}{3} = 7$

FIRST TERM, DIRECT RATIO, SECOND TERM, FRACTION FORM

❶ DIVIDE FIRST TERM BY SECOND TERM

① Divide 21 by 3 $(21 \div 3 = 7)$.

Inverse Ratio. An *inverse ratio* is a simple ratio that results when the second term is divided by the first term. See Figure 9-3. To avoid confusion, the inverse ratios are found using the ratio in fraction form. The fraction is inverted (reciprocal) when the numerator and denominator are interchanged.

INVERSE RATIO

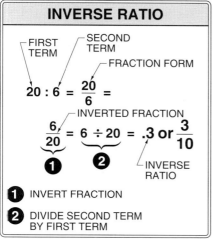

FIRST TERM — SECOND TERM — FRACTION FORM

$$20 : 6 = \frac{20}{6} =$$

INVERTED FRACTION

$$\frac{6}{20} = 6 \div 20 = .3 \text{ or } \frac{3}{10}$$

❶ ❷ — INVERSE RATIO

❶ INVERT FRACTION

❷ DIVIDE SECOND TERM BY FIRST TERM

Figure 9-3. An inverse ratio is the simple ratio that results when the second term is divided by the first term.

To find an inverse ratio, change the terms to a fraction, invert the fraction, and divide the new numerator by the new denominator.

For example, to find the inverse ratio of 20 : 6, change 20 : 6 to a fraction ($^{20}/_6$). Invert $^{20}/_6$ and divide 6 by 20 ($^6/_{20}$ = .3 or $^3/_{10}$).

EXAMPLE
Inverse Ratio

1. Find the inverse ratio of 36 to 9.

FIRST TERM — SECOND TERM — FRACTION FORM

$$36 : 9 = \frac{36}{9} =$$

INVERTED FRACTION

$$\frac{9}{36} = 9 \div 36 = .25 \text{ or } \frac{1}{4}$$

❶ ❷ — INVERSE RATIO

❶ INVERT FRACTION

❷ DIVIDE SECOND TERM BY FIRST TERM

① Invert $^3/_9$ to become $^9/_{36}$.
② Divide 9 by 36 ($^9/_{36}$ = 9 ÷ 36 = .25 or $^1/_4$).

Compound Ratio. A *compound ratio* is two ratio expressions in which the first terms are multiplied, the second terms are multiplied, and the relation of the first product to the second product is stated. See Figure 9-4.

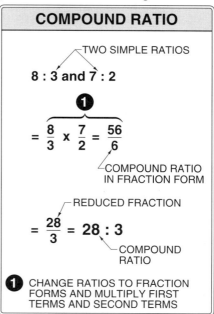

COMPOUND RATIO

TWO SIMPLE RATIOS

$$8 : 3 \text{ and } 7 : 2$$

❶

$$= \frac{8}{3} \times \frac{7}{2} = \frac{56}{6}$$

COMPOUND RATIO IN FRACTION FORM

REDUCED FRACTION

$$= \frac{28}{3} = 28 : 3$$

COMPOUND RATIO

❶ CHANGE RATIOS TO FRACTION FORMS AND MULTIPLY FIRST TERMS AND SECOND TERMS

Figure 9-4. To find a compound ratio of two simple ratios, change the ratios to fractions and multiply the numerators and the denominators.

To find a compound ratio of two simple ratios, change the ratios to fractions, multiply the numerators (first terms), and multiply the denominators (second terms). The product is a compound ratio in fraction form.

For example, to find the compound ratio of the ratios 8 : 3 and 7 : 2, change the ratios to fractions ($\frac{8}{3}$ and $\frac{7}{2}$). Multiply 8 by 7 and 3 by 2 ($\frac{8 \times 7}{3 \times 2} = \frac{56}{6} = \frac{28}{3}$). The compound ratio in fraction form is $\frac{56}{6}$ which reduces to $\frac{28}{3}$.

EXAMPLE
Compound Ratio

1. Find the compound ratio of 6 : 1 and 3 : 2.

TWO SIMPLE RATIOS

6 : 1 and 3 : 2

❶

$$= \frac{6}{1} \times \frac{3}{2} = \frac{18}{2}$$

COMPOUND RATIO IN FRACTION FORM

REDUCED FRACTION

$$= \frac{9}{1} = 9 : 1$$

COMPOUND RATIO

❶ CHANGE RATIOS TO FRACTION FORMS AND MULTIPLY FIRST TERMS AND SECOND TERMS

① Change the ratios to fractions to get $\frac{6}{1}$ and $\frac{3}{2}$. Multiply 6 by 3 and 1 by 2 ($\frac{6 \times 3}{1 \times 2} = \frac{18}{2} = \frac{9}{1}$). The answer in fraction form is $\frac{18}{2}$ which reduces to $\frac{9}{1}$. The fraction $\frac{9}{1}$ is then written as a compound ratio. The compound ratio of 6 : 1 and 6 : 2 = **9 : 1**.

Fractions in Ratios. A ratio expression composed of whole numbers, such as 8 : 4 or 9 : 12, can be expressed as $\frac{8}{4}$ or $\frac{9}{12}$. Both of these ratios are composed of whole numbers expressed in fraction form simply to indicate division.

A ratio expression can have a fraction in the first term, second term, or both terms. See Figure 9-5.

To solve fractions in ratios, change the whole number to a fraction, invert the divisor (second term), and multiply the first term by the second term.

For example, to solve 5 : $\frac{3}{4}$, change the whole number 5 to a fraction ($5 = \frac{5}{1}$) and divide by the second term $\frac{3}{4}$ ($\frac{5}{1} \div \frac{3}{4}$). To divide $\frac{5}{1}$ by $\frac{3}{4}$, invert $\frac{3}{4}$ to become $\frac{4}{3}$ and multiply ($\frac{5}{1} \times \frac{4}{3} = \frac{20}{3}$).

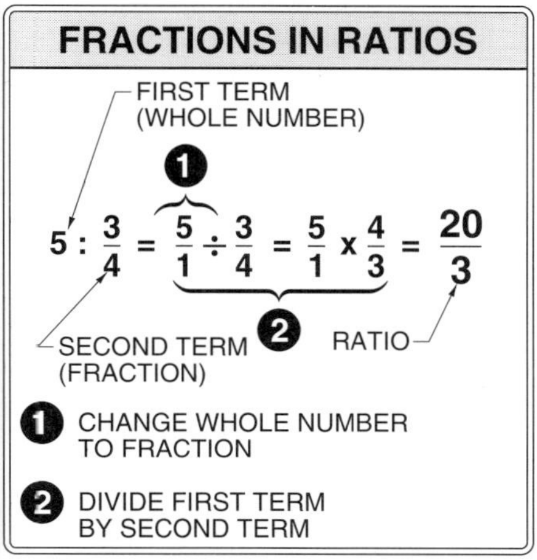

FRACTIONS IN RATIOS

FIRST TERM
(WHOLE NUMBER)

①

$$5 : \frac{3}{4} = \frac{5}{1} \div \frac{3}{4} = \frac{5}{1} \times \frac{4}{3} = \frac{20}{3}$$

SECOND TERM **②** RATIO
(FRACTION)

① CHANGE WHOLE NUMBER TO FRACTION

② DIVIDE FIRST TERM BY SECOND TERM

Figure 9-5. To solve fractions in ratios, change the whole number to a fraction and divide the first term by the second term.

EXAMPLES

Fractions in Ratios

1. Find the ratio of $\frac{3}{4}$ to 8.

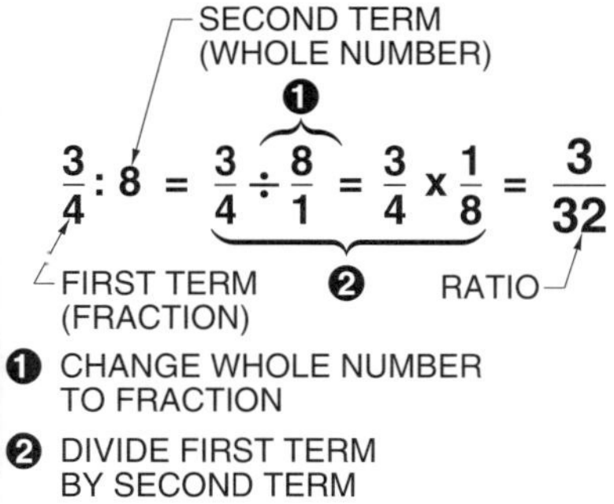

SECOND TERM
(WHOLE NUMBER)

①

$$\frac{3}{4} : 8 = \frac{3}{4} \div \frac{8}{1} = \frac{3}{4} \times \frac{1}{8} = \frac{3}{32}$$

FIRST TERM **②** RATIO
(FRACTION)

① CHANGE WHOLE NUMBER TO FRACTION

② DIVIDE FIRST TERM BY SECOND TERM

① Change 8 to a fraction. ② Divide $\frac{3}{4}$ by $\frac{8}{1}$ ($\frac{3}{4} \div \frac{8}{1} = \frac{3}{4} \times \frac{1}{8} = \frac{3}{32}$).

2. Find the ratio of $\frac{2}{5}$ to $\frac{2}{3}$.

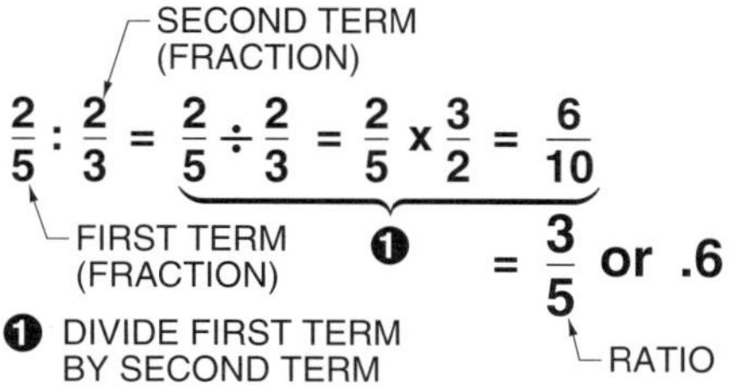

SECOND TERM
(FRACTION)

$$\frac{2}{5} : \frac{2}{3} = \frac{2}{5} \div \frac{2}{3} = \frac{2}{5} \times \frac{3}{2} = \frac{6}{10}$$

FIRST TERM
(FRACTION) **①** $= \frac{3}{5}$ or .6

① DIVIDE FIRST TERM BY SECOND TERM
RATIO

① Divide $\frac{2}{5}$ by $\frac{2}{3}$ ($\frac{2}{5} \div \frac{2}{3} = \frac{2}{5} \times \frac{3}{2} = \frac{6}{10} = \frac{3}{5}$ or **.6**).

Finding First Term. To find the first term when the second term and the ratio are known, multiply the second term by the ratio. See Figure 9-6.

For example, to find the first term (x) of $x : 3 = 12$, multiply the second term, 3, by 12 ($3 \times 12 = $ **36**). The complete ratio expression is **36 : 3 = 12**.

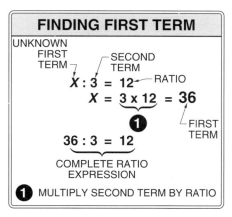

FINDING FIRST TERM

UNKNOWN FIRST TERM

SECOND TERM

$$X : 3 = 12 \quad \text{RATIO}$$

$$X = \underbrace{3 \times 12}_{} = 36$$

① FIRST TERM

$$\underbrace{36 : 3 = 12}_{}$$

COMPLETE RATIO EXPRESSION

① MULTIPLY SECOND TERM BY RATIO

Figure 9-6. To find the first term when the second term and ratio are known, multiply the second term by the ratio.

EXAMPLE

Finding First Term

1. Find the first term of $x : 6 = 5$.

UNKNOWN
FIRST TERM ⌐
⌐ SECOND TERM

$\overset{X}{} : 6 = 5$ ← RATIO

$X = \underbrace{6 \times 5}_{\textbf{❶}} = 30$

└ FIRST TERM

$\underbrace{30 : 6 = 5}$

COMPLETE RATIO
EXPRESSION

❶ MULTIPLY SECOND TERM BY RATIO

① Multiply the second term, 6, by 5 ($6 \times 5 = 30$). The complete ratio expression is **30 : 6 = 5**.

Finding Second Term. To find the second term when the first term and the ratio are known, divide the first term by the ratio. See Figure 9-7.

FINDING SECOND TERM

FIRST TERM ⌐
⌐ UNKNOWN SECOND TERM

$14 : \overset{X}{} = .7$ ← RATIO

$X = \underbrace{14 \div .7}_{\textbf{❶}} = 20$

└ SECOND TERM

$\underbrace{14 : 20 = .7}$

COMPLETE RATIO
EXPRESSION

❶ DIVIDE FIRST TERM BY RATIO

Figure 9-7. To find the second term when the first term and ratio are known, multiply the first term by the ratio.

For example, to find the second term (x) of $14 : x = .7$, divide the first term, 14, by .7 ($14 \div .7 = 20$). The complete ratio expression is **14 : 20 = .7**.

EXAMPLE

Finding Second Term

1. Find the second term of $35 : x = 7$.

FIRST TERM ⌐
⌐ UNKNOWN SECOND TERM

$35 : \overset{X}{} = 7$ ← RATIO

$X = \underbrace{35 \div 7}_{\textbf{❶}} = 5$

└ SECOND TERM

$\underbrace{35 : 5 = 7}$

COMPLETE RATIO
EXPRESSION

❶ DIVIDE FIRST TERM BY RATIO

① Divide 35 by 7 ($35 \div 7 = 5$).

PRACTICE PROBLEMS

Simple Ratios

1. Find the ratio of 36 to 4.
2. Find the compound ratio of 9 : 2 and 10 : 3.
3. Find the inverse ratio of 3 to 9.
4. Find the ratio of 7 to 49.
5. Find the ratio of 6 to 9.
6. Find the inverse ratio of $\frac{2}{3}$ to 5.
7. Find the compound ratio of 22 : 1 and 3 : 4.
8. Find the ratio of $6\frac{1}{2}$ to 78. *(3 places)*

9. Find the ratio of 16 to 66.
10. Find the ratio of 5 to 4.
11. Find the inverse ratio of $5\frac{3}{4}$ to $17\frac{1}{4}$.
12. Find the ratio of $3\frac{1}{3}$ to $16\frac{2}{3}$.
13. Find the inverse ratio of 24 to 48.
14. Find the ratio of 60 to $\frac{1}{2}$.
15. Find the compound ratio of 10 : 3 and 6 : 4.
16. Find the first term of x : 16 = 48.
17. Find the first term of x : 3.2 = 640.
18. Find the first term of x : 6 = 54.
19. Find the second term of 1.4 : x = 70.
20. Find the second term of 9 : x = 54.

Ratio Transformation

Ratio transformation is the change of the ratio form. In ratio expressions, a change of form through multiplying or dividing the first term and/or the second term has various effects on the ratio.

Multiplying or Dividing First Term. When the first term of a ratio expression is multiplied, the ratio is multiplied by the same amount. When the first term of a ratio expression is divided, the ratio is divided by the same amount. See Figure 9-8.

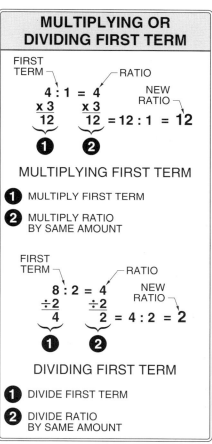

MULTIPLYING OR DIVIDING FIRST TERM

FIRST TERM — RATIO

4 : 1 = 4 NEW RATIO
x 3   x 3
12   12 = 12 : 1 = 12

1 **2**

MULTIPLYING FIRST TERM

1 MULTIPLY FIRST TERM

2 MULTIPLY RATIO BY SAME AMOUNT

FIRST TERM — RATIO

8 : 2 = 4 NEW RATIO
÷2   ÷2
4   2 = 4 : 2 = 2

1 **2**

DIVIDING FIRST TERM

1 DIVIDE FIRST TERM

2 DIVIDE RATIO BY SAME AMOUNT

Figure 9-8. When the first term is multiplied or divided, the ratio is multiplied or divided by the same amount.

For example, if the first term in 4 : 1 = 4 is multiplied by 3, the ratio is also multiplied by 3. Multiply the term 4 by 3 ($4 \times 3 = 12$) and multiply the ratio 4 by 3 ($4 \times 3 = 12$). The ratio 4 : 1 = 4 becomes **12 : 1 = 12**.

If the first term in 8 : 2 = 4 is divided by 2, then the ratio is divided by 4. Divide the term 8 by 4 ($8 \div 4 = 2$) and divide the ratio

4 by 2 (4 ÷ 2 = 2). The ratio 8 : 2 = 4 becomes **4 : 2 = 2**.

 EXAMPLES

Multiplying or Dividing First Term

1. Multiply the first term of 10 : 2 = 5 by 4.

FIRST
TERM ⌐   ⌐ RATIO
$$10 : 2 = 5$$
$$\underline{\times 4} \quad \underline{\times 4}$$
$$40 \quad 20 = 40 : 2 = 20$$
❶   **❷**

❶ MULTIPLY FIRST TERM
❷ MULTIPLY RATIO BY SAME AMOUNT

① Multiply the first term, 10, by 4 (10 × 4 = 40). ② Multiply the ratio, 5, by 4 (5 × 4 = 20). The ratio 10 : 2 = 5 becomes **40 : 2 = 20**.

2. Divide the first term of 90 : 3 = 30 by 3.

FIRST
TERM ⌐   ⌐ RATIO
$$90 : 3 = 30$$
$$\underline{\div 3} \quad \underline{\div 3}$$
$$30 \quad 10 = 30 : 3 = 10$$
❶   **❷**

❶ DIVIDE FIRST TERM
❷ DIVIDE RATIO BY SAME AMOUNT

① Divide the first term, 90, by 3 (90 ÷ 3 = 30). ② Divide the ratio by 3 (30 ÷ 3 = 10). The ratio 90 : 3 = 30 becomes **30 : 3 = 10**.

Multiplying or Dividing Second Term. When the second term of a ratio expression is multiplied, the ratio is divided by the same amount. When the second term of a ratio expression is divided, the ratio is multiplied by the same amount. See Figure 9-9.

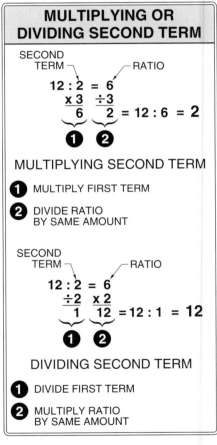

Figure 9-9. When the second term is multiplied or divided, the ratio is divided or multiplied by the same amount.

For example, if the second term in 12 : 2 = 6 is multiplied by 3, the ratio is divided by 3. Multiply the term 2 by 3 to get 6 and divide the ratio 6 by 3 to get 2. The ratio 12 : 2 = 6 becomes **12 : 6 = 2**.

If the second term in 12 : 2 = 6 is divided by 2, then the ratio is multiplied by 2. Divide the term 2 by 2 to get 1 and multiply the ratio 6 by 2 to get 12. The ratio 12 : 2 = 6 becomes **12 : 1 = 12**.

 EXAMPLES

Multiplying or Dividing Second Term

1. Multiply the second term of 24 : 2 = 12 by 4.

SECOND
TERM ⌐   ⌐RATIO
$$24 : 2 = 12$$
$$\quad \times 4 \quad \div 4$$
$$24 : 8 = 3$$
❸ ❶ ❷

❶ MULTIPLY SECOND TERM
❷ DIVIDE RATIO BY SAME AMOUNT
❸ CARRY FIRST TERM

① Multiply the second term, 2, by 4 (2 × 4 = 8). ② Divide the ratio, 12, by 4 (12 ÷ 4 = 3). ③ Carry the first term. The ratio 24 : 2 = 12 becomes **24 : 8 = 3**.

2. Divide the second term of 15 : 3 = 5 by 3.

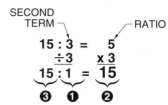

SECOND
TERM ⌐   ⌐RATIO
$$15 : 3 = 5$$
$$\quad \div 3 \quad \times 3$$
$$15 : 1 = 15$$
❸ ❶ ❷

❶ DIVIDE SECOND TERM
❷ MULTIPLY RATIO BY SAME AMOUNT
❸ CARRY FIRST TERM

① Divide the second term, 3, by 3 (3 ÷ 3 = 1). ② Multiply the ratio, 5, by 3 (5 × 3 = 15). ③ Carry the first term. The ratio 15 : 3 = 5 becomes **15 : 1 = 15**.

Multiplying or Dividing Both Terms. When both terms are multiplied or divided by the same number, the ratio is not altered. See Figure 9-10.

For example, if both terms in 8 : 4 = 2 are multiplied by 3, the ratio does not change. Multiply the first term, 8, by 3 (8 × 3 = 24). Multiply the second term, 4, by 3 (4 × 3 = 12). The ratio 8 : 4 = 2 becomes **24 : 12 = 2**.

If both terms in 15 : 10 = 1.5 are divided by the same number, the ratio is not altered. Divide the first term, 15, by 5 (15 ÷ 5 = 3). Divide the second term, 10, by 5 (10 ÷ 5 = 2). The ratio 15 : 10 = 1.5 becomes **3 : 2 = 1.5**.

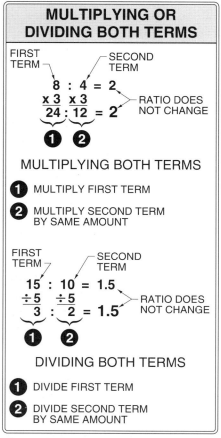

MULTIPLYING OR DIVIDING BOTH TERMS

FIRST TERM — — SECOND TERM

$$8 : 4 = 2$$
$$\text{x 3} \quad \text{x 3}$$
$$\overline{24 : 12} = 2 \quad \text{RATIO DOES NOT CHANGE}$$

❶ ❷

MULTIPLYING BOTH TERMS

❶ MULTIPLY FIRST TERM

❷ MULTIPLY SECOND TERM BY SAME AMOUNT

FIRST TERM — — SECOND TERM

$$15 : 10 = 1.5$$
$$\div 5 \quad \div 5$$
$$\overline{3 : 2} = 1.5 \quad \text{RATIO DOES NOT CHANGE}$$

❶ ❷

DIVIDING BOTH TERMS

❶ DIVIDE FIRST TERM

❷ DIVIDE SECOND TERM BY SAME AMOUNT

Figure 9-10. When both terms are multiplied or divided by the same number, the ratio is not altered.

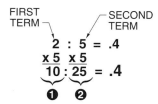

EXAMPLES

Multiplying or Dividing Both Terms

1. Multiply both terms of 2 : 5 = .4 by 5.

FIRST TERM — — SECOND TERM

$$2 : 5 = .4$$
$$\text{x 5} \quad \text{x 5}$$
$$\overline{10 : 25} = .4$$

❶ ❷

❶ MULTIPLY FIRST TERM

❷ MULTIPLY SECOND TERM BY SAME AMOUNT

① Multiply the first term, 2, by 5 (2 × 5 = 10). ② Multiply the second term, 5, by 5 (5 × 5 = 25). The ratio 2 : 5 = .4 becomes **10 : 25 = .4**.

2. Divide both terms of 12 : 3 = 4 by 3.

FIRST TERM — — SECOND TERM

$$12 : 3 = 4$$
$$\div 3 \quad \div 3$$
$$\overline{4 : 1} = 4$$

❶ ❷

❶ DIVIDE FIRST TERM

❷ DIVIDE SECOND TERM BY SAME AMOUNT

① Divide the first term, 12, by 3 (12 ÷ 3 = 4). ② Divide the second term, 3, by 3 (3 ÷ 3 = 1). The ratio 12 : 3 = 4 becomes **4 : 1 = 4**.

PRACTICE PROBLEMS

Ratio Transformation

1. Multiply the first term of 16 : 4 by 3.

2. Divide the first term of 14 : 7 by 3.5.

3. Multiply the second term of 8 : 2 = 4 by 4.
4. Divide the second term of 9 : 3 = 3 by .5.
5. Multiply both terms of 16 : 4 = 4 by ½.
6. Divide both terms of 20 : 5 = 4 by 2.5.

PROPORTION

A *proportion* is an expression of equality between two ratios. For example, the ratio expressions 8 : 4 and 12 : 6 both have a ratio of 2. All proportions are composed of two equal ratio expressions. See Figure 9-11.

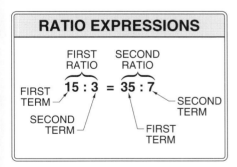

RATIO EXPRESSIONS

Figure 9-11. All proportions are composed of two equal ratio expressions.

The equal sign (=) is used between the two ratio expressions to indicate that one expression is equal to the other. Four dots (::) can be used in place of the equal sign,

but the equal sign is most common. The equal proportion is written as 8 : 4 = 12 : 6.

Proportion can also be expressed in fraction form. The ratio expressions 8 : 4 and 12 : 6 can be written ⁸⁄₄ and ¹²⁄₆. This proportion is read, "8 is to 4 as 12 is to 6."

The ratio to the left of the equal sign is the first ratio, and the ratio to the right is the second ratio. Each ratio has a first and second term.

For example, in the proportion 8 : 4 = 12 : 6, the 8 is the first term of the first ratio, the 4 is the second term of the first ratio, the 12 is the first term of the second ratio, and the 6 is the second term of the second ratio.

In a proportion where the first term is larger than the second term in the first ratio, the first term must also be larger than the second term in the second ratio. For example, in the proportion 8 : 2 = 12 : 3, the 8 and 12 (first terms) are larger than the 2 and 3 (second terms).

If the first term is smaller than the second term in the first ratio, the first term must also be smaller than the second term in the second ratio. For example, in the ratio 12 : 20 = 15 : 25 the 12 and 15 (first terms) are smaller than the 20 and 25 (second terms).

Means and Extremes

Means are the two inner numbers of a proportion. *Extremes* are the two outer numbers of a proportion. See Figure 9-12. For example, in the proportion 8 : 4 = 12 : 6 the 4 and 12 (two inner numbers) are the means. The 8 and 6 (two outer numbers) are the extremes.

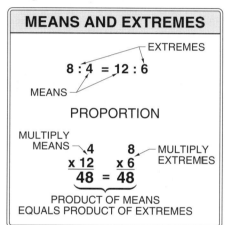

Figure 9-12. Means are the two inner numbers and extremes are the two outer numbers of a proportion.

The product of the means is equal to the product of the extremes. For example, in the proportion 8 : 4 = 12 : 6, multiply 4 by 12 (means) (4 × 12 = **48**) and multiply 8 by 6 (extremes) (8 × 6 = **48**).

The ratios are equal because the product of the means is equal to the product of the extremes. This is true in any proportion. No proportion expression is a true proportion unless the two ratios are equal.

Finding Missing Mean. The product of the extremes divided by either mean gives the other mean as the quotient. See Figure 9-13. For example, in the proportion 8 : x = 12 : 6, multiply 8 by 6 (extremes) (8 × 6 = 48). Divide 48 by 12 (known mean) (48 ÷ 12 = 4). Therefore, x = **4**.

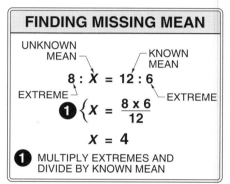

Figure 9-13. To find a missing mean, multiply the extremes and divide by the known mean.

EXAMPLES

Finding Missing Mean

1. Find the missing mean in 5 : .3 = x : 6.

① Multiply 5 by 6 (extremes) (5 × 6 = 30) and divide 30 by .3 (known mean) (30 ÷ .3 = 100). Therefore, $x = $ **100**.

2. Find the missing mean in 36 : x = 12 : 17.

UNKNOWN
MEAN ⌐ KNOWN
MEAN

$$36 : X = 12 : 17$$

EXTREME ⌐ EXTREME

$$❶ \left\{ X = \frac{36 \times 17}{12} \right.$$

$$X = 51$$

❶ MULTIPLY EXTREMES AND DIVIDE BY KNOWN MEAN

① Multiply 36 by 17 (36 × 17 = 612) and divide 612 by 12 (612 ÷ 12 = 51). Therefore, $x = $ **51**.

Finding Missing Extreme. The product of the means divided by either extreme gives the other extreme as the quotient. See Figure 9-14.

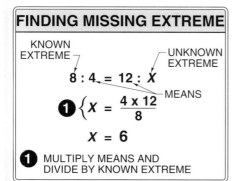

FINDING MISSING EXTREME

KNOWN
EXTREME ⌐ UNKNOWN
EXTREME

$$8 : 4 = 12 : X$$

MEANS

$$❶ \left\{ X = \frac{4 \times 12}{8} \right.$$

$$X = 6$$

❶ MULTIPLY MEANS AND DIVIDE BY KNOWN EXTREME

Figure 9-14. To find a missing extreme, multiply the means and divide by the known extreme.

For example, in the proportion 8 : 4 = 12 : x, multiply 4 by 12 (means) (4 × 12 = 48) and divide 48 by 8 (known extreme) (48 ÷ 8 = 6). Therefore, $x = $ **6**.

EXAMPLES

Finding Missing Extreme

1. Find the missing extreme in 3 : 4 = 12 : x.

KNOWN
EXTREME ⌐ UNKNOWN
EXTREME

$$3 : 4 = 12 : X$$

MEANS

$$❶ \left\{ X = \frac{4 \times 12}{3} \right.$$

$$X = 16$$

❶ MULTIPLY MEANS AND DIVIDE BY KNOWN EXTREME

① Multiply 4 by 12 (means) (4 × 12 = 48) and divide 48 by 3 (known extreme) (48 ÷ 3 = 16). Therefore, $x = $ **16**.

2. Find the missing extreme in x : 120 = 8 : 192.

UNKNOWN
EXTREME ⌐ KNOWN
EXTREME

$$X : 120 = 8 : 192$$

MEANS

$$❶ \left\{ X = \frac{120 \times 8}{192} \right.$$

$$X = 5$$

❶ MULTIPLY MEANS AND DIVIDE BY KNOWN EXTREME

① Multiply 120 by 8 (120 × 8 = 960) and divide 960 by 192 (960 ÷ 192 = 5). Therefore, $x = 5$.

PRACTICE PROBLEMS

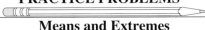

Means and Extremes

Find the missing number (x).

1. $17 : x = 51 : 54$
2. $\frac{19}{x} = \frac{209}{143}$
3. $x : 3.9 = 40 : 78$
4. $\frac{6}{.5} = \frac{x}{20}$
5. $\frac{35}{7} = \frac{x}{91}$
6. $48 : 20 = x : 50$
7. $x : 300 = 20 : 100$
8. $1 : x = 7 : 84$
9. $48 : x = 67.25 : 201.75$
10. $12 : 5 = x : 40$
11. $16 : 24 = x : 15$
12. $4 : 4\frac{2}{3} = 9\frac{3}{7} : x$
13. $48 : x = 240 : 6$
14. $x : 9 = 13 : 10$
15. $2.76 : 3.45 = 2.28 : x$

Applying Proportions

Proportions are either direct proportions or inverse proportions. A *direct proportion* is a statement of equality between two ratios in which the first of four terms divided by the second equals the third divided by the fourth.

For example, in the proportion $8 : 4 = 12 : 6$, both ratios equal 2 ($8 ÷ 4 = 2$, $12 ÷ 6 = 2$). An increase in one term results in a proportional increase in the other related term.

For example, in the proportion $8 : 4 = 12 : 6$, if the 8 is increased, then the 12 must be proportionally increased by the same amount to keep the proportion equal.

An *inverse proportion* is the opposite of direct proportion. An increase in one quantity results in a proportional decrease in the other related quantity. See Figure 9-15.

For example, in a two-pulley system, if one pulley is 10″ in diameter and the other is 2″ in diameter, the smaller pulley rotates five times when the larger pulley rotates one time ($10 ÷ 2 = 5$). This results in $10 : 2 = 1 : 5$, which is a false proportion because the ratios are not equal.

To make it a true inverse proportion, the second ratio is inverted to get $10 : 2 = 5 : 1$. The number of rotations of the pulley is inversely proportional to the diameter of the pulley.

If the diameter of the small pulley is increased to 5″, then it rotates twice as the large pulley rotates once ($10 : 5 = 2 : 1$). The diameter of the small pulley is increased, and the number of rotations inversely decreased.

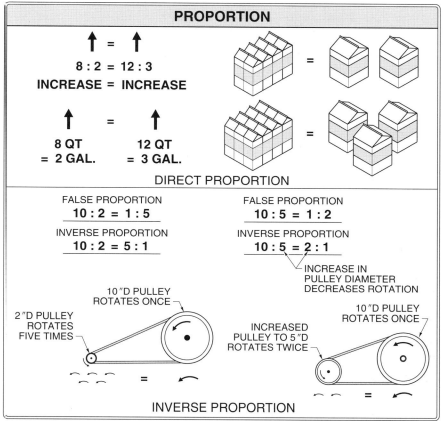

Figure 9-15. In a direct proportion, an increase in one quantity results in a proportional increase in the other related quantity. In an inverse proportion, an increase in one quantity results in a proportional decrease in the other related quantity.

Applications of proportions are presented as problems that give a statement of certain conditions and an indication of the required answer. In all proportion problems, three of the four numbers, which become terms of the two ratios, are given in the statement of the problem.

First it must be determined if the problem is a direct or inverse pro-portion, then the unknown number is found. See Figure 9-16.

Three terms are given in the problem, "If a metal joint 16′ long requires 80 rivets, how many rivets are required for a joint 11′ long?" To solve the problem, determine which of the quantities or numbers is the first term in the first ratio, which is the second term in the first ratio, etc. Remember that the first two terms in a proportion form a ratio, the last two terms form a ratio, and a ratio can only be made of numbers with common measures.

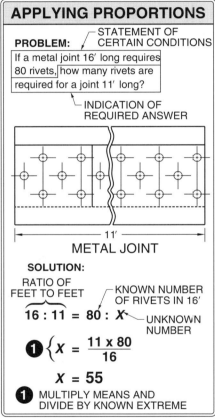

APPLYING PROPORTIONS

PROBLEM: ⌐ STATEMENT OF CERTAIN CONDITIONS

If a metal joint 16' long requires 80 rivets, how many rivets are required for a joint 11' long?

└ INDICATION OF REQUIRED ANSWER

METAL JOINT

|← 11' →|

SOLUTION:

RATIO OF FEET TO FEET ⌐ KNOWN NUMBER OF RIVETS IN 16'

$16 : 11 = 80 : X$ — UNKNOWN NUMBER

❶ $\left\{ X = \dfrac{11 \times 80}{16} \right.$

$X = 55$

❶ MULTIPLY MEANS AND DIVIDE BY KNOWN EXTREME

Figure 9-16. Proportions are presented as problems that give a statement of certain conditions and an indication of the required answer.

The first ratio in the proportion consists of 16 and 11 which are in feet. This is a comparison of one measurement in feet to another measurement in feet. The second ratio must also have numbers representing common units of measure. The number 80 represents rivets and the unknown number also represents rivets. Therefore, the 80 and the unknown number form the second ratio.

To decide where to place the three numbers given in any problem, always make the unknown number the second term in the second ratio. The unknown number is indicated by the lowercase letter x. The first terms of both ratios are the terms that complete a statement.

For example, "a metal joint 16' long requires 80 rivets" is a complete statement, and 16 and 80 are the first terms of the ratios. The second term of the first ratio is 11'. Thus, the ratio is $16 : 11 = 80 : x$ because terms with common units of measure are compared in ratios. If this were an inverse proportion, the second ratio would be inverted to get $16 : 11 = x : 80$.

In a direct proportion, the second term of the second ratio is an extreme of a proportion. To solve for the unknown number of a direct proportion, find the missing extreme by multiplying 11 and 80 ($11 \times 80 = 880$) and divide 880 by 16 ($880 \div 16 = 55$). Therefore, the number of rivets in a joint 11' long is **55**.

In an inverse proportion, the first term of the second ratio is a mean of a proportion. To solve for the unknown number of an inverse proportion, find the missing mean.

EXAMPLES

| **Applying Proportions** |

1. If 15 ironworkers can build a form 12′ high in one day, how many ironworkers are required to build a form 20′ high in one day?

KNOWN EXTREME ⌐   ⌐ UNKNOWN EXTREME

$$12 : 20 = 15 : X$$

MEANS

❶ $\left\{ X = \dfrac{20 \times 15}{12} \right.$

$$X = 25$$

❶ MULTIPLY MEANS AND DIVIDE BY KNOWN EXTREME

This example is a direct proportion. Set up the proportion with the *x* (unknown number of ironworkers) as the second term of the second ratio, 15 (number of ironworkers) as the first term of the second ratio, 12 (feet) as the first term of the first ratio (completes statement with 15 ironworkers), and 20 (feet) as the second term of the first ratio to get 12 : 20 = 15 : *x*.

① Multiply 20 by 15 (means) (20 × 15 = 300) and divide 300 by 12 (known extreme) (300 ÷ 12 = 25). Therefore, *x* = **25**.

2. The forces are inversely proportional to the distances the forces are applied when levers are used for mechanical work. Find the weight (force) that can be lifted by the lever.

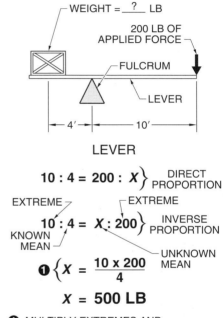

LEVER

$$10 : 4 = 200 : X \Big\} \quad \text{DIRECT PROPORTION}$$

EXTREME ⌐   ⌐ EXTREME

$$10 : 4 = X : 200 \Big\} \quad \text{INVERSE PROPORTION}$$

KNOWN MEAN ⌐   UNKNOWN MEAN

❶ $\left\{ X = \dfrac{10 \times 200}{4} \right.$

$$X = 500 \text{ LB}$$

❶ MULTIPLY EXTREMES AND DIVIDE BY KNOWN MEAN

Set up the proportion with the *x* (unknown weight) as the second term of the second ratio, 200 (pounds of force) as the first term of the second ratio, 10 (length from fulcrum to applied force) as the first term of the first ratio (completes statement with 200 lb), and 4 (length from fulcrum to weight) as the second term of the first ratio.

Because forces are inversely proportional to the distances that the forces are applied, invert the second ratio to get $10 : 4 = x : 200$.

① Multiply 10 by 200 (extremes) ($10 \times 200 = 2000$) and divide 2000 by 4 (known mean) ($2000 \div 4 = 500$). The lever can lift a weight of **500 lb**.

PRACTICE PROBLEMS

Applying Proportions

1. If 5 t of coal costs $150.00, how much does 3 t cost?
2. If 12 yd of cloth cost $48.00, how much does 4 yd cost?
3. What will 9 lb of a product cost if 3 lb costs $3.50?
4. If 3 qt of lemonade mix can be bought for $1.05, how many quarts can be bought for $3.15?
5. If a tour group travelled 150 mi in 3 hr, how far would they travel, at the same speed, in 7 hr?
6. How many hours would be required for a freight train to travel 180 mi if it were going the same speed as a train that travelled 240 mi in 6 hr?
7. If a tourist spent $320.00 during the first four days of vacation, how many days vacation could be taken for $800.00 at the same rate of expense?
8. If the taxes on a house with an assessed value of $120,000.00 were $3260.00, how much would the taxes be, at the same rate, on a house which was assessed at $180,000.00?
9. If 15 gal. of gasoline were used to drive 255 mi, how much gasoline would be used for a trip of 425 mi at the same rate?
10. If a printing press printed 675 pages of material in 15 min, how much time would be required to print 2700 pages at the same rate?

Compound Proportions

A *compound proportion* is a proportion where some of the terms are products of two variables. A *variable* is a quantity that may change. See Figure 9-17. For example, "If five welders fabricate 130 joint assemblies in 12 hours, then 10 welders can fabricate 195 joint assemblies in 9 hours" is a statement of a compound proportion.

To simplify the procedure of solving compound proportions, each term of the ratios is renamed as either a cause or an effect. Every question in any proportion, especially in a compound proportion, may be considered as a comparison between two causes and two effects.

Causes are actions that produce effects. *Effects* are the results of actions or causes. For example, if a carpenter builds a deck, the action of building is the cause and the deck is the effect.

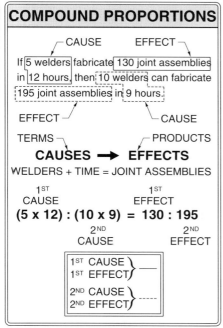

Figure 9-17. In a compound proportion, some of the terms are products of two variables.

In the statement, "If five welders fabricate 130 joint assemblies in 12 hours, then 10 welders can fabricate 195 joint assemblies in 9 hours," the welders (5 and 10) and the time (12 hr and 9 hr) are the causes, and the number of joint assemblies (130 and 195) are the effects.

To simplify the causes, multiply the related terms. The proportion is written $(5 \times 12) : (10 \times 9) = 130$: 195. When using the words *causes* and *effects*, the proportion (5×12) : $(10 \times 9) = 130 : 195$ is written in 1st cause : 2nd cause = 1st effect : 2nd effect form. The 5×12 is the 1st cause, the 10×9 is the 2nd cause, the 130 is the 1st effect, and the 195 is the 2nd effect.

The unknown term (answer required) in a proportion is the second term of the second ratio. See Figure 9-18. When solving compound proportion problems, the unknown term can be a cause or an effect.

If the unknown term is an effect, write the compound proportion in 1st cause : 2nd cause = 1st effect : 2nd effect form. If the unknown term is a cause, write the proportion in 1st effect : 2nd effect = 1st cause : 2nd cause form. In both cases the unknown term is the second term of the second ratio.

Consider the problem, "If 18 workers build a wall 420′ long in 16 days, how many workers can build a wall 280′ long in eight days?" The "workers" and "days" are the causes, and "feet" are the effects.

The workers build the wall over a number of days, and the number of feet the wall is built is an effect, or the result of action. The unknown term is the number of workers. Effects (feet) form the first ratio, and causes (workers and days) form the second ratio.

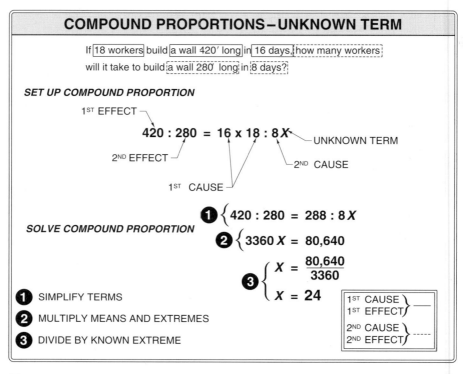

Figure 9-18. The unknown term in a proportion is the second term of the second ratio.

To determine the number of workers, set up the compound proportion by placing an x as the second term of the second ratio. The first term of the second ratio must be workers because ratios only include common measures.

Since causes are workers and days, the terms of the second ratio must be a product of days and workers (16×18 and $8 \times x$). Place 420 (feet related to the 1st cause) as the 1st effect, and 280 (feet related to the 2nd cause) as the 2nd effect.

Multiply 16 (days) by 18 (workers) ($16 \times 18 = 288$) (1st cause) and 8 (days) by x (workers) ($8 \times x = 8x$) (2nd cause). Multiply 280 by 288 (means) to get 80,640. Divide 80,640 by 3360 (known extreme) to get 24. Thus, **24** workers are needed to build a wall 280' long in eight days.

EXAMPLE

Compound Proportions

1. If it takes four days for 100 workers to erect one story of a building, how many stories can 50 workers erect in 64 days?

SET UP COMPOUND PROPORTION

FIRST CAUSE — SECOND EFFECT —

4 x 100 : 64 x 50 = 1 : X

SECOND CAUSE — FIRST EFFECT

SOLVE COMPOUND PROPORTION

❶ { 400 : 3200 = 1: X

❷ { 400 X = 3200

❸ { $X = \dfrac{3200}{400}$

$X = 8$

❶ SIMPLIFY TERMS
❷ MULTIPLY MEANS AND EXTREMES
❸ DIVIDE BY KNOWN EXTREME

The unknown term is work accomplished, or an effect. The 1st and 2nd causes in the first ratio are workers and days because the x (work) is an effect and is the second term of the second ratio.

The first term of the second ratio must be work because ratios only include common measures. Thus, the 1st cause in the second ratio is one story. The causes are the terms of workers and days, so the terms of the second ratio must be a product of workers and days.

Place 100 and 4 (workers and days related to the 1st effect) as the 1st cause, and 50 and 64 (workers and days related to the 2nd effect) as the 2nd effect. ① Multiply 100 (workers) by 4 (days) to get 400 (1st cause) and multiply 50 (workers) by 64 (days) to get 3200 (2nd cause). Thus, the proportion is 400 : 3200 = 1 : x. ② Multiply 3200 by 1 (means) to get 3200 and 400 by x (extremes) to get 400x. ③ Divide 3200 by 400 (known extreme) to get 8. Thus, **8 stories** are erected by 50 workers in 64 days.

PRACTICE PROBLEMS

Compound Proportions

1. If 15 workers excavate 240 cu ft in four days, how many days are required for 50 workers to excavate 1000 cu ft?

2. If a person travels 120 mi in three 12-hr days, how many 9-hr days are required to travel 360 mi?

3. If 16 horses consume 128 bushels of oats in 50 days, how many bushels will five horses consume in 90 days?

4. If six workers dig a ditch 34 yd long in 10 days, how many yards can 20 workers dig in 15 days?

5. If 20 masons lay 35,000 bricks in three days, how many bricks can 15 masons lay in eight days?

6. If one electrician can wire three rooms in 6 hr, how many rooms can two electricians wire in 9 hr?

7. If two cabinetmakers spray and finish 12 cabinets in 4 hr, how many cabinets can one cabinetmaker spray and finish in 6 hr?

8. If two carpenters build 12 trusses in 8 hr, how many trusses can six carpenters build in 16 hr?

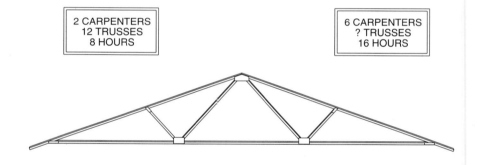

```
┌─────────────────┐          ┌─────────────────┐
│  2 CARPENTERS   │          │  6 CARPENTERS   │
│   12 TRUSSES    │          │   ? TRUSSES     │
│    8 HOURS      │          │    16 HOURS     │
└─────────────────┘          └─────────────────┘
```

Name _____ Date _____

True-False and Completion

_____ **1.** _____ is the relationship between two quantities or terms.

_____ **2.** The _____ is the symbol used to indicate a relation between terms.

_____ **3.** A(n) _____ ratio is a simple ratio which results when the first term is divided by the second term.

T F **4.** To find the ratio of two terms, divide the first term by the second term.

_____ **5.** A(n) _____ ratio is a simple ratio which results when the second term is divided by the first term.

_____ **6.** Ratio _____ is the change of the ratio form.

_____ **7.** A(n) _____ is an expression of equality between two ratios.

_____ **8.** _____ are the two inner numbers of a proportion.

_____ **9.** _____ are the two outer numbers of a proportion.

_____ **10.** A(n) _____ proportion is a proportion where some of the terms are products of two variables.

_____ **11.** _____ are the results of actions or causes.

T F **12.** In a proportion, the product of the means is twice the product of the extremes.

T F **13.** A simple ratio is the ratio between two terms.

T F **14.** When both terms of a ratio are multiplied or divided by the same number, the ratio is not altered.

T F **15.** No proportion expression is a true proportion unless the two ratios are equal.

Calculations

_____ **1.** Find the ratio of 24 to 4.

_____ **2.** Find the inverse ratio of 32 to 8.

_____ **3.** Find the compound ratio of 3 : 1 and 5 : 3.

_____ **4.** Find the ratio of $\frac{1}{4}$ to 16.

_____ **5.** Find the first term of $x : 4 = 3$.

_____ **6.** Find the second term of $42 : x = 14$.

_____ **7.** What is the inverse ratio of 64 to 16?

_____ **8.** What is the first term of $x : 6 = .5$?

_____ **9.** What is the second term of $128 : x = 4$?

_____ **10.** What is the new ratio of 4 : 1 if the first term is multiplied by 3?

_____ **11.** Find the missing mean in $6 : .5 = x : 5$.

_____ **12.** Find the missing extreme in $3 : 4 = 16 : x$.

_____ **13.** If three workers can make 120 hamburgers in 1 hr, how many hamburgers can four workers make in 3 hr?

_____ **14.** If a tourist travels 300 mi in 10 hr, how many hours are required to travel 60 mi?

Name _____ Date _____

_____ **1.** Find the ratio of 44 to 11.

_____ **2.** Find the ratio of .50 to .10.

_____ **3.** Find the inverse ratio of 36 to 3.

_____ **4.** Find the inverse ratio of .75 to .25.

_____ **5.** Find the compound ratio of 5 : 1 and 5 : 4.

_____ **6.** Find the compound ratio of 6 : 1 and 7 : 1.

_____ **7.** Find the first term of $x : 16 = .4$.

_____ **8.** What is the first term of $x : 8 = .25$.

_____ **9.** What is the second term of $12 : x = .3$.

_____ **10.** What is the new ratio of 5 : 1 if the first term is multiplied by 4?

_____ **11.** If a tractor-trailer truck travels 460 mi in 9 hr, how many hours are required to travel 520 mi?

233

_____ **12.** If one laborer mops 100 sq ft of two-ply fibrous asphalt felt on a foundation wall in 3.4 hr, how many square feet can three laborers mop in 4 hr?

_____ **13.** If a paperhanger hangs 400 sq ft of vinyl wall covering in 7 hr, how many hours are required to cover a 12′ × 20′ room with 8′ walls? Disregard openings. *Round the answer to the nearest whole hour.*

_____ **14.** A laborer breaks up 1 cu yd of medium density rock in 1.6 hr. How many cubic yards can be broken up by three laborers during an 8-hr shift?

_____ **15.** Two tractor operators backfill a foundation with 100 cu yd of earth in .9 hr. In how many hours can the two tractor operators backfill 800 cu yd of earth?

_____ **16.** If an electrician pulls 100′ of No. 14 Cu conductor in .7 hr, how many feet can be pulled in 4 hr?

_____ **17.** A plumber installs 100′ of 4″ Schedule 40 black pipe in 6 hr. How many hours are required to install 240′?

PLANE FIGURES

A formula is a mathematical equation that contains a fact, rule, or principle. Formulas are used in all trade areas. They are used to find the area of plane figures such as circles, triangles, quadrilaterals, etc.

FORMULAS

An *equation* is a means of showing that two numbers or two groups of numbers are equal to the same amount. See Figure 10-1.

For example, a baker has two pies of the same size. Pie A is cut into six equal pieces, and Pie B is cut into eight equal pieces. One customer buys three pieces of Pie A, and another customer buys four pieces of Pie B. The customers buy the same amount of pie because $\frac{3}{6} = \frac{1}{2}$ and $\frac{4}{8} = \frac{1}{2}$. All equations must balance, and as $\frac{3}{6}$ and $\frac{4}{8}$ each equal $\frac{1}{2}$, $\frac{3}{6} = \frac{4}{8}$ is an equation.

A *formula* is a mathematical equation that contains a fact, rule, or principle. Italic letters are used in formulas to represent values (amounts).

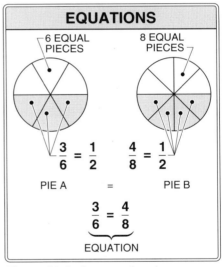

Figure 10-1. An equation is a means of showing that two numbers, or two groups of numbers, are equal to the same amount.

For example, $a + b = c$ is a formula. See Figure 10-2. In a formula, any number or letter may be transposed from the left to the right or from the right to the left of the equal sign. When transposed, the sign of the number or letter is changed to the opposite sign.

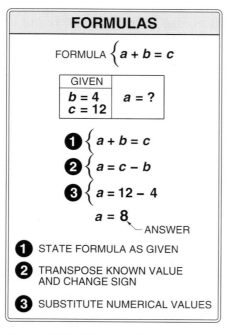

FORMULAS

FORMULA $\{ a + b = c$

GIVEN	
$b = 4$	$a = ?$
$c = 12$	

❶ $\{ a + b = c$

❷ $\{ a = c - b$

❸ $\{ a = 12 - 4$

$a = 8$

ANSWER

❶ STATE FORMULA AS GIVEN

❷ TRANSPOSE KNOWN VALUE AND CHANGE SIGN

❸ SUBSTITUTE NUMERICAL VALUES

Figure 10-2. A formula is a mathematical equation that contains a fact, rule, or principle.

The sign is always in front of the number or letter of which it is a part. For example, in the formula $a + b = c$, if $b = 4$ and $c = 12$, the value of a is found by changing the formula to $a = c - b$, or $a = 12 - 4$. A formula can be changed to solve for any unknown value if the other values are known.

Common Formulas

In the trades, common formulas related to plane and solid figures are used when laying out jobs. For example, a welder may be required to lay out and build a cylindrical tank to hold a specified number of gallons of liquid. By applying the volume formula for cylinders, the welder can determine the required size of the cylindrical tank.

PLANE FIGURES

A *plane figure* is a flat figure with no depth. All plane figures are composed of straight or curved lines. Plane figures include circles, triangles, quadrilaterals, and polygons. The area of a plane figure is measured in square units such as square inches, square feet, square millimeters, square meters, etc.

Lines

A *line* is the boundary of a surface. See Figure 10-3. Lines are measured in linear units such as inches, feet, yards, millimeters, meters, etc.

A *straight line* is the shortest distance between two points. It is commonly referred to as a line. A *curved line* is a line that continually changes direction. It is commonly referred to as a curve.

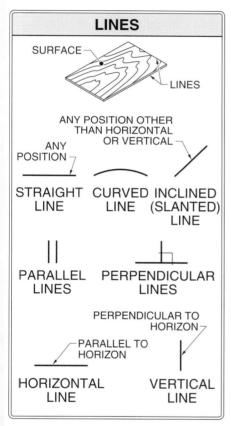

Figure 10-3. A line is the boundary of a surface.

A *horizontal line* is a line that is parallel to the horizon. It may be referred to as a level line. A *perpendicular line* is a line that makes a 90° angle with another line. The symbol for perpendicular is ⊥.

A *vertical line* is a line that is perpendicular to the horizon. It is often referred to as a plumb line. *Plumb* is an exact verticality (determined by a plumb bob and line) with the Earth's surface.

All lines may be drawn in any position, unless they are horizontal or vertical. An *inclined line* is a line that is slanted. It is neither horizontal nor vertical. *Parallel lines* are two or more lines that remain the same distance apart. The symbol for parallel lines is ∥.

Angles

An *angle* is the intersection of two lines or sides. See Figure 10-4. The *vertex* is the point of intersection of the sides of an angle. To identify angles, letters are placed at the end of each side and at the vertex.

When referring to an angle, the vertex letter is read second. The angle symbol (∠) is used to indicate an angle. The way the two sides of an angle intersect determines the size and type of angle.

Angles are measured in degrees, minutes, and seconds. The symbol for degrees is °, the symbol for minutes is ′, the symbol for seconds is ″. There are 360° in a circle (one revolution). There are 60′ in one degree and 60″ in one minute. For example, an angle might contain 112°-30′-12″.

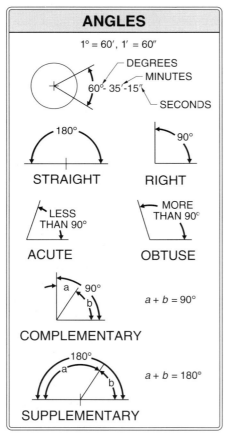

Figure 10-4. An angle is the intersection of two lines.

A *straight angle* is two lines that intersect to form a straight line. It is one-half of a revolution, or $360/_2 = 180°$. A straight angle always contains 180°.

A *right angle* is two lines that intersect perpendicular to each other. It is one-fourth of a revolution, or $360/_4 = 90°$. A right angle always contains 90°.

An *acute angle* is an angle that contains less than 90°. An *obtuse angle* is an angle that contains more than 90°. For example, a 45° angle is an acute angle, and a 135° angle is an obtuse angle.

Lines can intersect to create more than one angle. *Complementary angles* are two angles formed by three lines in which the sum of the two angles equals 90°. Each complementary angle is an acute angle. For example, a 30° angle and a 60° angle are acute angles that are complementary angles.

To find the complementary angle of a known acute angle, subtract the known angle from 90. For example, to find the complementary angle of a 40° angle, subtract 40 from 90 (90 − 40 = 50). The complementary angle to a 40° angle is a 50° angle.

Supplementary angles are two angles formed by three lines in which the sum of the two angles equals 180°. For example, a 45° angle and a 135° angle are supplementary angles.

To find the supplementary angle of a known angle, subtract the known angle from 180. For example, to find the supplementary angle of a 70° angle, subtract 70 from 180 (180 − 70 = 110). The supplementary angle to a 70° angle is a 110° angle.

Adjacent angles are angles that have the same vertex and one side in common. Adjacent angles are formed when two or more lines intersect.

For example, when two straight lines intersect, four angles and four sets of adjacent angles are formed. The sum of adjacent angles that form a straight line equals 180°. The two angles opposite each other when two straight lines intersect are equal.

Area

Area is the number of unit squares equal to the surface of an object. For example, a standard size sheet of plywood is 4′ × 8′. It contains an area of 32 sq ft (4 × 8 = 32 sq ft).

Area is expressed in square inches, square feet, and other units of measure. A *square inch* measures 1″ × 1″ or its equivalent. A *square foot* contains 144 sq in. (12″ × 12″ = 144 sq in.). The area of any plane figure can be determined by applying the proper formula. See Figure 10-5.

Circles

A *circle* is a plane figure generated about a centerpoint. See Figure 10-6. All circles contain 360°. The *circumference* is the boundary of a circle.

The *diameter* is the distance from circumference to circumference through the centerpoint. The *centerpoint* is the point a circle or arc is drawn around.

An *arc* is a portion of the circumference. The *radius* is the distance from the centerpoint to the circumference. It is one-half the length of the diameter.

A *chord* is a line from circumference to circumference not through the centerpoint. A *quadrant* is one-fourth of a circle containing 90°.

A *sector* is a pie-shaped piece of a circle. A *segment* is the portion of a circle set off by a chord. A *semicircle* is one-half of a circle containing 180°.

Concentric circles are two or more circles with different diameters but the same centerpoint. *Eccentric circles* are two or more circles with different diameters and different centerpoints.

A *tangent* is a straight line touching the curve of the circumference at only one point. A tangent is perpendicular to the radius. A *secant* is a straight line touching the circumference at two points.

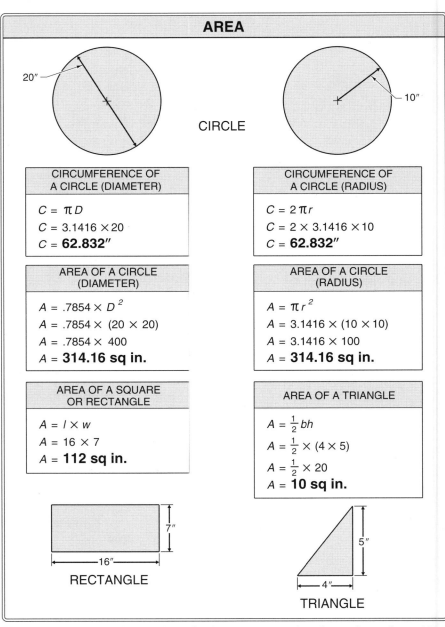

AREA

CIRCLE

20″   10″

CIRCUMFERENCE OF A CIRCLE (DIAMETER)
$C = \pi D$
$C = 3.1416 \times 20$
$C = \mathbf{62.832''}$

CIRCUMFERENCE OF A CIRCLE (RADIUS)
$C = 2\pi r$
$C = 2 \times 3.1416 \times 10$
$C = \mathbf{62.832''}$

AREA OF A CIRCLE (DIAMETER)
$A = .7854 \times D^2$
$A = .7854 \times (20 \times 20)$
$A = .7854 \times 400$
$A = \mathbf{314.16\ sq\ in.}$

AREA OF A CIRCLE (RADIUS)
$A = \pi r^2$
$A = 3.1416 \times (10 \times 10)$
$A = 3.1416 \times 100$
$A = \mathbf{314.16\ sq\ in.}$

AREA OF A SQUARE OR RECTANGLE
$A = l \times w$
$A = 16 \times 7$
$A = \mathbf{112\ sq\ in.}$

AREA OF A TRIANGLE
$A = \frac{1}{2}bh$
$A = \frac{1}{2} \times (4 \times 5)$
$A = \frac{1}{2} \times 20$
$A = \mathbf{10\ sq\ in.}$

7″

◄───16″───►

RECTANGLE

5″

◄─4″─►

TRIANGLE

Figure 10-5. Area is the number of unit squares equal to the surface of an object.

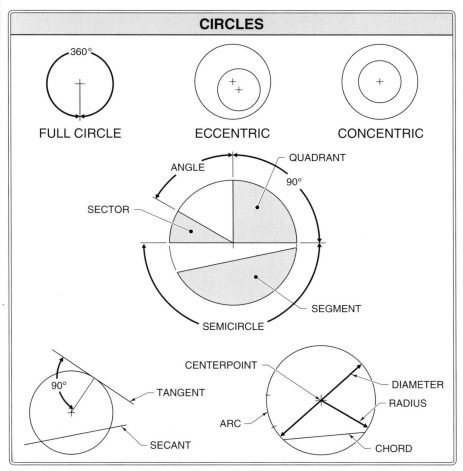

Figure 10-6. A circle is a plane figure generated about a centerpoint.

Circumference of a Circle (Diameter). When the diameter is known, the circumference of a circle is found by applying the formula:

$$C = \pi D$$

where

C = circumference

π = 3.1416

D = diameter

For example, what is the circumference of a 20″ diameter circle?

$$C = \pi D$$

$$C = 3.1416 \times 20$$

$$C = \textbf{62.832″}$$

Circumference of a Circle (Radius). When the radius is known, the circumference of a circle is found by applying the formula:

$$C = 2\pi r$$

where

C = circumference

2 = constant

π = 3.1416

r = radius

For example, what is the circumference of a 10″ radius circle?

$$C = 2\pi r$$

$$C = 2 \times 3.1416 \times 10$$

$$C = \textbf{62.832″}$$

Area of a Circle (Diameter). When the diameter is known, the area of a circle is found by applying the formula:

$$A = .7854 \times D^2$$

where

A = area

$.7854$ = constant

D^2 = diameter squared

For example, what is the area of a 28″ diameter circle?

$$A = .7854 \times D^2$$

$$A = .7854 \times (28 \times 28)$$

$$A = .7854 \times 784$$

$$A = \textbf{615.754 sq in.}$$

Area of a Circle (Radius). When the radius is known, the area of a circle is found by applying the formula:

$$A = \pi r^2$$

where

A = area

π = 3.1416

r^2 = radius squared

For example, what is the area of a 14″ radius circle?

$$A = \pi r^2$$

$$A = 3.1416 \times (14 \times 14)$$

$$A = 3.1416 \times 196$$

$$A = \textbf{615.754 sq in.}$$

PRACTICE PROBLEMS

Circles

Unless otherwise indicated, decimals are rounded to two places (hundredths).

1. What is the circumference of a 15″ diameter circle?
2. What is the circumference of a 4″ radius circle?
3. Find the area of a 21″ diameter circle.
4. Find the area of an 11″ radius circle.
5. What is the circumference of Circle A?
6. What is the area of Circle A?
7. What is the circumference of Circle B?

8. What is the area of Circle B?

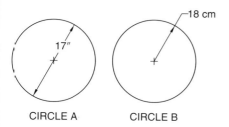

CIRCLE A   CIRCLE B

Triangles

A *triangle* is a three-sided polygon with three interior angles. The sum of the three angles of a triangle is always 180°. The sign (Δ) indicates a triangle. See Figure 10-7.

The *altitude* of a triangle is the perpendicular dimension from the vertex to the base. The *base* of a triangle is the side upon which the triangle stands. Any side can be taken as the base.

The angles of a triangle are named by uppercase letters. The sides of a triangle are named by lowercase letters. For example, a triangle may be named ΔABC and contain sides d, e, and f.

The different kinds of triangles are right triangles, isosceles triangles, equilateral triangles, and scalene triangles.

A *right triangle* is a triangle that contains one 90° angle and no equal sides. An *isosceles triangle* is a triangle that contains two equal angles and two equal sides.

An *equilateral triangle* is a triangle that has three equal angles and three equal sides. Each angle of an equilateral triangle is 60°.

A *scalene triangle* is a triangle that has no equal angles or equal sides. A scalene triangle may be acute or obtuse. An *acute triangle* is a scalene triangle with each angle less than 90°. An *obtuse triangle* is a scalene triangle with one angle greater than 90°.

Area of a Triangle. The area of a triangle is found by applying the formula:

$$A = \frac{1}{2}bh$$

where

A = area

$\frac{1}{2}$ = constant

b = base

h = height

For example, what is the area of a triangle with a 10″ base and a 12″ height?

$$A = \frac{1}{2}bh$$

$$A = \frac{1}{2} \times (10 \times 12)$$

$$A = \frac{1}{2} \times 120$$

$$A = \textbf{60 sq in.}$$

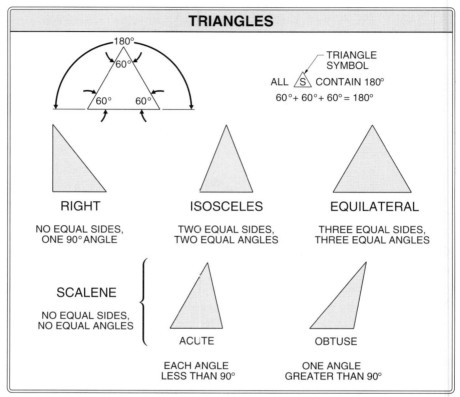

Figure 10-7. A triangle is a three-sided polygon with three interior angles.

Pythagorean Theorem. The *Pythagorean Theorem* states that the square of the hypotenuse of a right triangle is equal to the sum of the squares of the other two sides. The *hypotenuse* is the side of a right triangle opposite the right angle.

A right triangle is said to have a 3-4-5 relationship and often is used for laying out right angles and checking corners for squareness. To check a corner for squareness, measure 3′ along one side and 4′ along the other side. These two points measure 5′ apart when the corner is square. See Figure 10-8.

The length of the hypotenuse of a right triangle is found by applying the formula:

$$c = \sqrt{a^2 + b^2}$$

where

c = length of hypotenuse

a^2 = length of one side squared

b^2 = length of other side squared

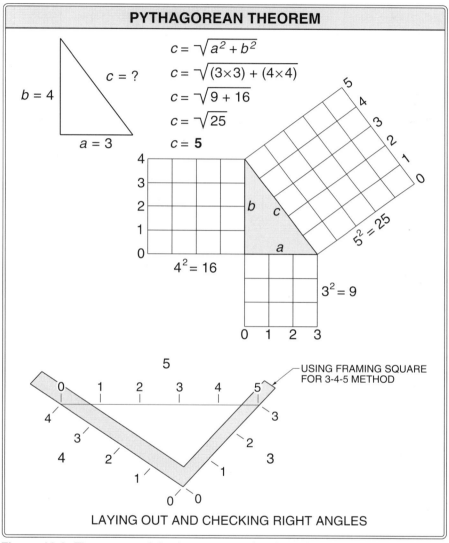

Figure 10-8. The square of the hypotenuse of a right triangle is equal to the sum of the squares of the other two sides of the triangle.

For example, what is the length of the hypotenuse of a triangle having sides of 3′ and 4′?

$$c = \sqrt{a^2 + b^2}$$

$$c = \sqrt{(3 \times 3) + (4 \times 4)}$$

$$c = \sqrt{9 + 16}$$

$$c = \sqrt{25}$$

$$c = \mathbf{5'}$$

PRACTICE PROBLEMS

Triangles

1. What is the area of a triangle with a 16″ base and a 22″ height?
2. What is the area of a triangle having an 8″ base and 3.5″ height?
3. What is the length of the hypotenuse of a triangle having an 8″ base and a 12″ height?
4. What is the area of Triangle A?
5. What is the length of the hypotenuse of Triangle A?
6. What is the area of Triangle B?
7. What is the length of the hypotenuse of Triangle B?

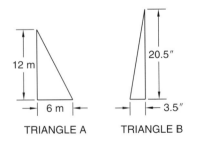

TRIANGLE A   TRIANGLE B

Regular Polygons

A *polygon* is a many-sided plane figure. All polygons are bound by straight lines. A *regular polygon* is a polygon with equal sides and equal angles. An *irregular polygon* has unequal sides and unequal angles. Polygons are named according to their number of sides. Typical polygons include the triangle (three sides), quadrilateral (four sides), pentagon (five sides), hexagon (six sides), and octagon (eight sides). See Figure 10-9.

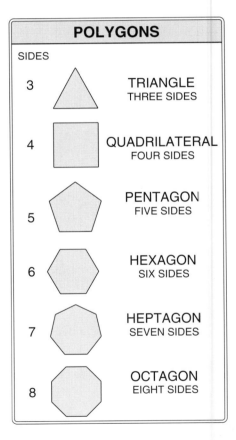

Figure 10-9. A polygon is a many-sided plane figure.

Quadrilaterals

A *quadrilateral* is a four-sided poly-gcn with four interior angles. The sum of the four angles of a quad-rilateral is always 360°. The kinds of quadrilaterals are squares, rec-tangles, rhombuses, rhomboids, trapezoids, and trapeziums. See Figure 10-10.

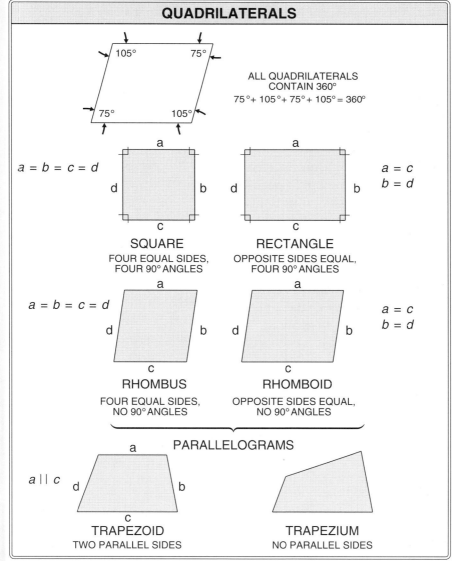

Figure 10-10. A quadrilateral is a four-sided polygon with four interior angles.

A *square* is a quadrilateral with all sides equal and four 90° angles. A *rectangle* is a quadrilateral with opposite sides equal and four 90° angles. A *rhombus* is a quadrilateral with all sides equal and no 90° angles. A *rhomboid* is a quadrilateral with opposite sides equal and no 90° angles.

The square, rectangle, rhombus, and rhomboid are parallelograms. A *parallelogram* is a four-sided plane figure with opposite sides parallel and equal.

A *trapezoid* is a quadrilateral with two sides parallel. A *trapezium* is a quadrilateral with no sides parallel. Trapezoids and trapeziums are not parallelograms because all opposite sides are not parallel.

Area of a Square or Rectangle. The area of a square or the area of a rectangle is found by applying the formula:

$$A = l \times w$$

where

A = area

l = length

w = width

For example, what is the area of a 22'-0" × 16'-0" storage room?

$$A = l \times w$$

$$A = 22 \times 16$$

$$A = \textbf{352 sq ft}$$

PRACTICE PROBLEMS

Quadrilaterals

1. What is the area of a 14'-6" × 16'-0" bedroom?
2. What is the area in square yards of a piece of 12' × 18' carpet?
3. How many square inches are in a square having 2.75" sides? *(4 places)*
4. A contractor has 3½ sheets of 4' × 8' plywood. How many square feet of plywood are there?
5. What is the area of Lot 312?
6. Lot 312 contains how many more square feet than Lot 313?

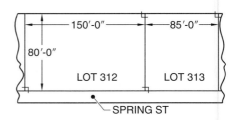

Name _____ Date _____

True-False and Completion

_____ **1.** A(n) _____ is a means of showing that two numbers or two groups of numbers are equal to the same amount.

_____ **2.** _____ is the number of unit squares equal to the surface of an object.

_____ **3.** The _____ Theorem states that the square of the hypotenuse of a right triangle is equal to the sum of the squares of the other two sides.

_____ **4.** A(n) _____ is a mathematical equation that contains a fact, rule, or principle.

_____ **5.** The _____ is the side of a right triangle opposite the right angle.

T   F   **6.** A vertical line is a line that is parallel to the horizon.

T   F   **7.** A line is the boundary of a surface.

T   F   **8.** An angle is the intersection of two lines.

_____ **9.** The _____ is the point of intersection of the sides of an angle.

_____ **10.** A(n) _____ figure is a flat figure that has no depth.

_____ **11.** In the formula $a + b = c$, what is the value of a if $b = 3$ and $c = 8$?

_____ **12.** _____ angles are two angles formed by three lines, in which the sum of the two angles equals $180°$.

_____ **13.** A(n) _____ is a portion of the circumference of a circle.

_____ **14.** _____ circles have different diameters and the same centerpoint.

_____ **15.** A(n) _____ is a pie-shaped piece of a circle.

_____ **16.** The _____ of a triangle is the perpendicular dimension from the vertex to the base.

_____ **17.** A(n) _____ triangle is a triangle that has no equal angles or equal sides.

_____ **18.** A(n) _____ is a many-sided plane figure.

T   F   **19.** A square is a quadrilateral with all sides equal and four 90° angles.

T   F   **20.** An isosceles triangle contains two equal angles and two equal sides.

T   F   **21.** The angles of a triangle are named by lowercase letters.

T   F   **22.** A quadrant is one-fourth of a circle.

T   F   **23.** A square foot contains 148 sq in.

T   F   **24.** A right angle always contains 90°.

T   F   **25.** Parallel lines always remain the same distance apart.

_____ **26.** The _____ is the distance from circumference to circumference through the centerpoint of a circle.

_____ **27.** A(n) _____ is a quadrilateral with all sides equal and no 90° angles.

_____ **28.** The sum of the four angles of a quadrilateral is always _____°.

_____ **29.** The radius of a circle is one-half the length of the _____.

_____ **30.** A(n) _____ line always makes a 90° angle with another line.

These formulas may be used to solve the following problems.

Circumference of a Circle (Diameter)...............$C = \pi D$

Circumference of a Circle (Radius).................$C = 2\pi r$

Area of a Circle (Diameter)..................$A = .7854 \times D^2$

Area of a Circle (Radius).....................................$A = \pi r^2$

Area of a Square or Rectangle ....$A = l \times w$

Area of a Triangle..$A = \frac{1}{2}bh$

Pythagorean Theorem................................$c = \sqrt{a^2 + b^2}$

Calculations

_____ **1.** A bedroom measures 14'-0" × 16'-0". How many square feet does the bedroom contain?

_____ **2.** A piece of steel plate measures 2.25" × 6.50". How many square inches does the plate contain?

_____ **3.** What is the area of a 10" radius circle?

_____ **4.** What is the area of a 24" diameter circle?

_____ **5.** What is the area of a triangle with an 8" base and a 12" height?

_____ **6.** What is the length of the hypotenuse of a triangle having sides of 4' and 6'?

_____ **7.** A field measures 125 m by 300 m. What is the area of the field?

    8. What is the area of a 15 cm radius circle?

    9. What is the area of a 25 cm diameter circle?

    10. What is the circumference of a 12″ diameter circle?

    11. What is the circumference of an 8″ radius circle?

    12. How many square feet does a 125′ × 205′ building lot contain?

    13. A patio measures 12′-6″ × 18′-0″. How many square feet does the patio contain?

    14. A triangle has a 10″ base and a 22″ height. What is the length of the hypotenuse?

    15. What is the circumference of a 14″ diameter circle?

    16. Which has the larger area, a 14″ diameter circle or a 14″ square?

    17. What is the length of the hypotenuse of a triangle having sides of 5′ and 8′?

    18. What is the area of a triangle with a 14 m base and a 22 m height?

Name _____ Date _____

These formulas may be used to solve the following problems.

Circumference of a Circle (Diameter)...............$C = \pi D$

Circumference of a Circle (Radius)..................$C = 2\pi r$

Area of a Circle (Diameter).................$A = .7854 \times D^2$

Area of a Circle (Radius)....................................$A = \pi r^2$

Area of a Square or Rectangle ....$A = l \times w$

Area of a Triangle...$A = \frac{1}{2}bh$

Pythagorean Theorem..................................$c = \sqrt{a^2 + b^2}$

---

1. A rectangular field measures 120′ on two sides and 65′ on the other two sides. How many square feet does the field contain?

2. What is the circumference of a pipe 36″ in diameter?

3. A circle has a radius of 15 cm. What is the area of the circle?

4. What is the area of a 12.25″ square?

5. What is the area of a right triangle with a base of 12″ and an altitude of 18″?

_____ 6. What is the length of the hypotenuse of a right triangle with a base of 8″ and an altitude of 12″?

_____ 7. What is the area of a triangle having 3′, 4′, and 5′ sides?

_____ 8. A field measures 145 m × 250 m. What is the area of the field?

_____ 9. How many square feet does a 150.25′ × 320.5′ lot contain?

_____ 10. Concentric circles have diameters of 12″ and 18″. What is the difference in area?

_____ 11. Which contains more area, an 11′ × 18′ rectangle or a circle 15′ in diameter?

_____ 12. What is the area of a triangle with a 12″ base and a 10″ height?

_____ 13. In a right triangle, side a is 6″ and side b is 9″. What is the length of side c? _(4 places)._

_____ 14. What is the area of a 16′-3″ × 18′-0″ bedroom?

_____ 15. A truck tire has a 26″ diameter. How many feet will the tire travel in three revolutions?

SOLID FIGURES

Solid figures have length, height, and depth. Solids are bound by plane surfaces or their surfaces are generated about an axis. Volume is the three-dimensional size of an object measured in cubic units.

SOLIDS

Polyhedra are solids bound by plane surfaces (faces). *Regular solids* (polyhedra) are solids with faces that are regular polygons (equal sides). *Irregular polyhedra* are solids with faces that are irregular polygons (unequal sides).

Solids have length, height, and depth. The five regular solids are the tetrahedron, hexahedron, octahedron, dodecahedron, and icosahedron. Other common solids are prisms, cylinders, pyramids, cones, and spheres. Less common solids include the torus and ellipsoid. See Figure 11-1.

Regular Solids

A *tetrahedron* is a regular solid of four triangles. A *hexahedron* is a regular solid of six squares. It is commonly referred to as a cube. An *octahedron* is a regular solid of eight triangles. A *dodecahedron* is a regular solid of twelve pentagons. An *icosahedron* is a regular solid of twenty triangles.

Prisms

A *prism* is a solid with two bases that are parallel and identical polygons. *Bases* are the ends of a prism. The three or more sides of a prism are parallelograms. See Figure 11-2. A prism can be triangular, rectangular, pentagonal, hexagonal, octagonal, etc. according to the shape of its bases.

Lateral faces are the sides of a prism. There are as many of these lateral faces as there are sides in one of the bases.

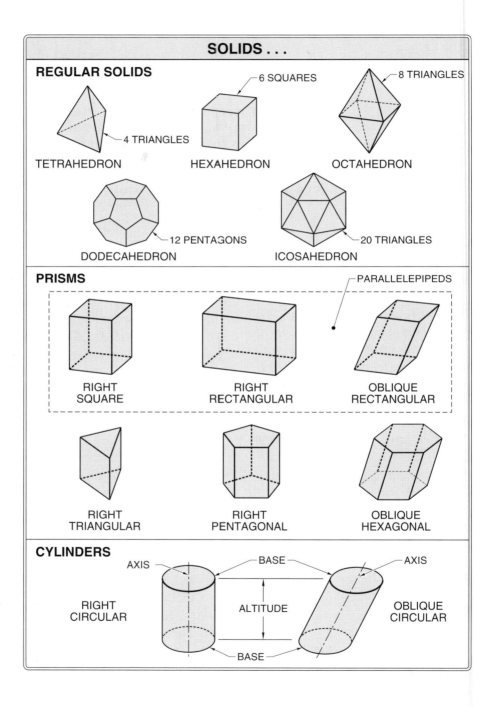

SOLIDS . . .

REGULAR SOLIDS

6 SQUARES

8 TRIANGLES

4 TRIANGLES

TETRAHEDRON   HEXAHEDRON   OCTAHEDRON

12 PENTAGONS

20 TRIANGLES

DODECAHEDRON   ICOSAHEDRON

PRISMS

PARALLELEPIPEDS

RIGHT
SQUARE

RIGHT
RECTANGULAR

OBLIQUE
RECTANGULAR

RIGHT
TRIANGULAR

RIGHT
PENTAGONAL

OBLIQUE
HEXAGONAL

CYLINDERS

AXIS   BASE   AXIS

RIGHT
CIRCULAR

ALTITUDE

OBLIQUE
CIRCULAR

BASE

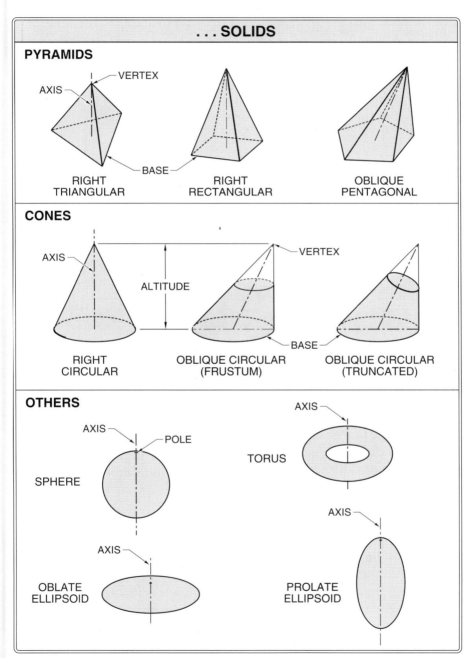

Figure 11-1. Solids have length, height, and depth.

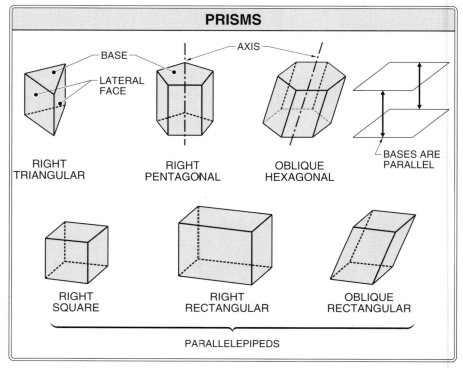

Figure 11-2. A prism is a solid with two bases that are parallel and identical polygons.

The *altitude* of a prism is the perpendicular distance between the two bases. When the bases are perpendicular to the faces, the altitude equals the edge of a lateral face.

A *right prism* is a prism with lateral faces perpendicular to the bases. An *oblique prism* is a prism with lateral faces not perpendicular to the bases.

A *parallelepiped* is a prism with bases that are parallelograms. A *right parallelepiped* is a prism with all edges perpendicular

to the bases. A *rectangular parallelepiped* is a prism with bases and faces that are all rectangles.

Cylinders

A *cylinder* is a solid generated by a straight line (genatrix) moving in contact with a curve and remaining parallel to the axis and its previous position. Each position of the genatrix forms an element of the cylinder.

A *right cylinder* is a cylinder with the axis perpendicular to the base. An *oblique cylinder* is a cylinder with the axis not perpendicular to the base. See Figure 11-3.

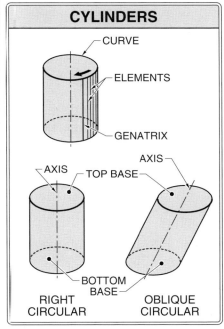

Figure 11-3. A cylinder is a solid generated by a straight line (genatrix) moving in contact with a curve and remaining parallel.

Pyramids

A *pyramid* is a solid with a base that is a polygon and sides that are triangles. The *vertex* is the common point of the triangular sides that forms the pyramid. See Figure 11-4.

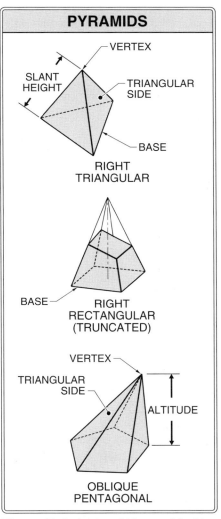

Figure 11-4. A pyramid is a solid with a base that is a polygon and sides that are triangles.

The *altitude* of a pyramid is the perpendicular distance from the vertex to the base. Pyramids are named according to the kind of polygon forming the base, such as triangular, quadrangular, pentagonal, and hexagonal.

A *regular pyramid* has a base that is a regular polygon and a vertex that is perpendicular to the center of the base. The *slant height* is the distance from the base to the vertex parallel to a side. It is the altitude of one of the triangles that forms the sides.

Cones

A *cone* is a solid generated by a straight line moving in contact with a curve and passing through the vertex. Cones have a circular base and a surface that tapers from the base to the vertex.

The altitude of a cone is the perpendicular distance from the vertex to the base. The slant height is the distance from the vertex to any point on the circumference of the base. See Figure 11-5.

Conic Sections. A *conic section* is a curve produced by a plane intersecting a right circular cone. A *right circular cone* is a cone with the axis at a 90° angle to the circular base. The four conic sections are the circle, ellipse, parabola, and hyperbola. See Figure 11-6.

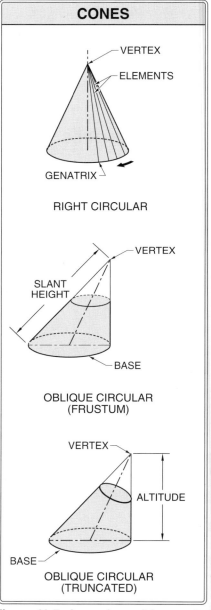

CONES

RIGHT CIRCULAR

OBLIQUE CIRCULAR
(FRUSTUM)

OBLIQUE CIRCULAR
(TRUNCATED)

Figure 11-5. A cone is a solid generated by a straight line (genatrix) moving in contact with a circle and passing through the vertex.

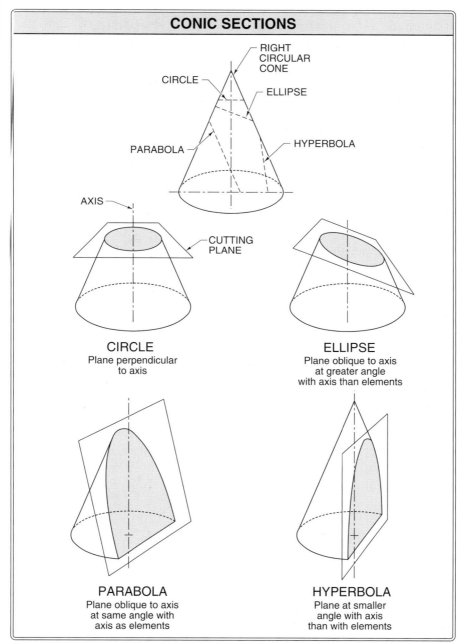

CONIC SECTIONS

RIGHT CIRCULAR CONE

CIRCLE

ELLIPSE

PARABOLA

HYPERBOLA

AXIS

CUTTING PLANE

CIRCLE
Plane perpendicular to axis

ELLIPSE
Plane oblique to axis at greater angle with axis than elements

PARABOLA
Plane oblique to axis at same angle with axis as elements

HYPERBOLA
Plane at smaller angle with axis than with elements

Figure 11-6. A conic section is a curve produced by a plane intersecting a right circular cone.

A *circle* is a plane figure formed by a cutting plane perpendicular to the axis of a cone. An *ellipse* is a plane figure formed by a cutting plane oblique to the axis of a cone, but at a greater angle with the axis than with the elements of the cone.

A *parabola* is a plane figure formed by a cutting plane oblique to the axis and parallel to the elements of the cone. A *hyperbola* is a plane figure formed by a cutting plane that has a smaller angle with the axis than with the elements of the cone.

Frustums. A *frustum* of a pyramid or cone is the remaining portion of a pyramid or cone with a cutting plane passed parallel to the base. A truncated pyramid or cone is the remaining portion of a pyramid or cone with the cutting plane passed not parallel to the base. See Figure 11-7.

Spheres

A *sphere* is a solid generated by a circle revolving about one of its axes. All points on the surface are an equal distance from the center of the sphere. See Figure 11-8.

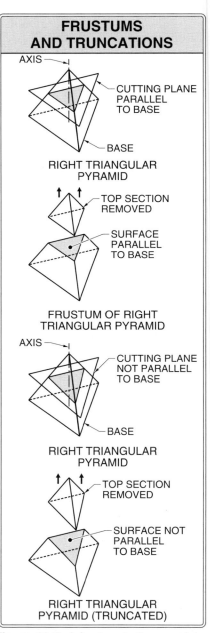

FRUSTUMS AND TRUNCATIONS

Figure 11-7. A frustum is the remaining piece of a pyramid or cone with a cutting plane passed parallel to the base.

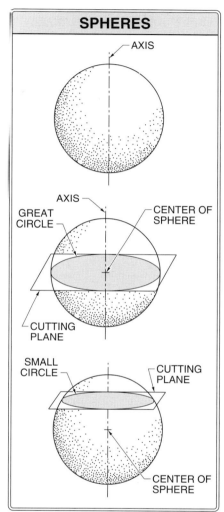

SPHERES

AXIS

AXIS

GREAT CIRCLE

CENTER OF SPHERE

CUTTING PLANE

SMALL CIRCLE

CUTTING PLANE

CENTER OF SPHERE

Figure 11-8. A sphere is a solid generated by a circle revolving about the axis.

A *great circle* is the circle formed by passing a cutting plane through the center of a sphere. A *small circle* is the circle formed by passing a cutting plane through a sphere but not through the center.

The circumference of a sphere is equal to the circumference of a great circle.

VOLUME

Volume is the three-dimensional size of an object measured in cubic units. For example, the volume of a standard size concrete block is 1024 cu in. (8″ × 8″ × 16″ = 1024 cu in.).

Volume is expressed in cubic inches, cubic feet, cubic yards, and other units of cubic measure. A *cubic inch* measures 1″ × 1″ × 1″ or its equivalent.

A *cubic foot* contains 1728 cu in. (12″ × 12″ × 12″ = 1728 cu in.). A *cubic yard* contains 27 cu ft (3′ × 3′ × 3′ = 27 cu ft).

Finding Volume

The volume of a solid figure can be determined by applying the proper formula. See Figure 11-9.

Volume of a Rectangular Solid. The volume of a rectangular solid is found by applying the formula:

$$V = l \times w \times h$$

where

V = volume

l = length

w = width

h = height

For example, what is the volume of a 24″ × 12″ × 8″ rectangular solid?

$V = l \times w \times h$

$V = 24 \times 12 \times 8$

$V =$ **2304 cu in.**

Volume of a Cylinder (Diameter).
When the diameter is known, the volume of a cylinder is found by applying the formula:

$V = .7854 \times D^2 \times h$

where

V = volume

$.7854$ = constant

D^2 = diameter squared

h = height

VOLUME . . .

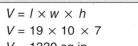
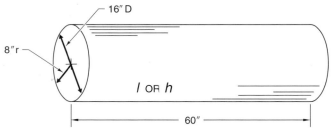

VOLUME OF A RECTANGULAR SOLID
$V = l \times w \times h$
$V = 19 \times 10 \times 7$
$V = 1330$ sq in.
$V = \frac{1330}{144} =$ **9.236 sq ft**

VOLUME OF A CYLINDER (DIAMETER)	VOLUME OF A CYLINDER (RADIUS)
$V = .7854 \times D^2 \times h$	$V = \pi r^2 \times l$
$V = .7854 \times (16 \times 16) \times 60$	$V = 3.1416 \times (8 \times 8) \times 60$
$V = .7854 \times 256 \times 60$	$V = 3.1416 \times 64 \times 60$
$V = 12{,}063.744$ cu in.	$V = 12{,}063.744$ cu in.
$V = \frac{12{,}063.744}{1728}$	$V = \frac{12{,}063.744}{1728}$
$V =$ **6.981$\overline{3}$ cu ft**	$V =$ **6.981$\overline{3}$ cu ft**

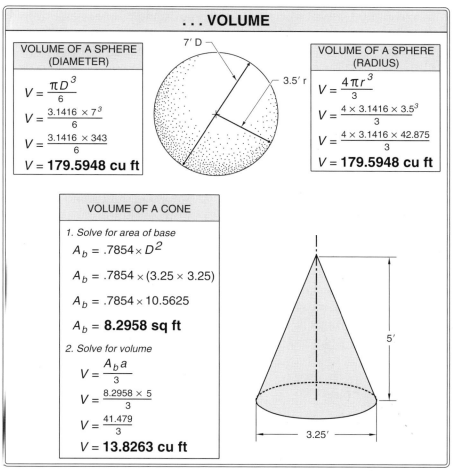

Figure 11-9. Volume is the three-dimensional size of an object.

For example, what is the volume of a tank that is 4'-0" in diameter and 12'-0" long? (*3 places*)

$$V = .7854 \times D^2 \times h$$

$$V = .7854 \times (4 \times 4) \times 12$$

$$V = .7854 \times 16 \times 12$$

$$V = 150.797 \text{ cu ft}$$

Volume of a Cylinder (Radius).
When the radius is known, the volume of a cylinder is found by applying the formula:

$$V = \pi r^2 \times l$$

where

V = volume

π = 3.1416

r^2 = radius squared

l = length

For example, what is the volume of a tank that has a 2'-0" radius and is 12'-0" long? *(3 places)*

$V = \pi r^2 \times l$

$V = 3.1416 \times (2 \times 2) \times 12$

$V = $ **150.797 cu ft**

Volume of a Cone. The volume of a cone is found by first solving for the area of the base and then solving for volume. The area of the base is found by applying the formula:

$A_b = .7854 \times D^2$

where

A_b = area of base

.7854 = constant

D^2 = diameter squared

The volume of the cone is then found by applying the formula:

$V = \dfrac{A_b a}{3}$

where

V = volume

A_b = area of base

a = altitude

3 = constant

For example, what is the volume of a cone that has a 14" diameter and a 35" altitude? *(3 places)*

1. *Solve for the area of the base.*

$A_b = .7854 \times D^2$

$A_b = .7854 \times (14 \times 14)$

$A_b = .7854 \times 196$

$A_b = $ **153.938 sq in.**

2. *Solve for the volume.*

$V = \dfrac{A_b a}{3}$

$V = \dfrac{153.938 \times 35}{3}$

$V = \dfrac{5387.83}{3}$

$V = $ **1795.94$\overline{3}$ cu in.**

Volume of a Sphere (Diameter). When the diameter is known, the volume of a sphere is found by applying the formula:

$V = \dfrac{\pi D^3}{6}$

where

V = volume

π = 3.1416

D^3 = diameter cubed

6 = constant

For example, what is the volume of a sphere that is 4'-0" in diameter? *(4 places)*

$$V = \frac{\pi D^3}{6}$$

$$V = \frac{3.1416 \times 4^3}{6}$$

$$V = \frac{3.1416 \times 64}{6}$$

$$V = \textbf{33.5104 cu ft}$$

Volume of a Sphere (Radius).

When the radius is known, the volume of a sphere is found by applying the formula:

$$V = \frac{4\pi r^3}{3}$$

where

V = volume

4 = constant

π = 3.1416

r^3 = radius cubed

3 = constant

For example, what is the volume of a sphere that has a 2'-0" radius? (4 places)

$$V = \frac{4\pi r^3}{3}$$

$$V = \frac{4 \times 3.1416 \times 8}{3}$$

$$V = \textbf{33.5104 cu ft}$$

PRACTICE PROBLEMS

Solids

1. What is the volume of a 4" × 8" × 36" rectangular solid?
2. What is the volume of a 3.5" × 6.25" × 18" rectangular solid?
3. What is the volume of a tank that has a 1.75' radius and is 6' long?
4. How many cubic feet of liquid will the Storage Drum hold? *Round the answer to the nearest cubic foot.*

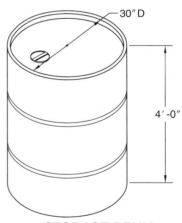

30" D

4'-0"

STORAGE DRUM

5. What is the volume of a sphere that has an 11" diameter?
6. Sphere A has a 22" diameter. Sphere B has a 16" diameter. How much larger is the volume of Sphere A?

7. How many cubic inches are in a cone that has a base diameter of 2″ and an altitude of 9″?

8. What is the volume of a sphere that has an 11″ radius?

9. Which of the three Cartons will hold the most material?

10. A ditch is dug 6′ deep, 4′ wide, and 32′ long. How many cubic yards of earth are removed from the ditch? *Round the answer to the nearest cubic yard.*

11. Which of the three Pipes will hold the least liquid?

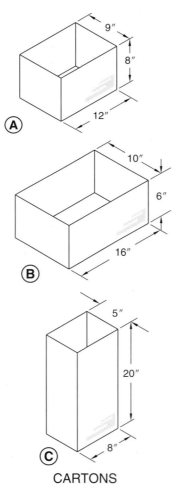

CARTONS

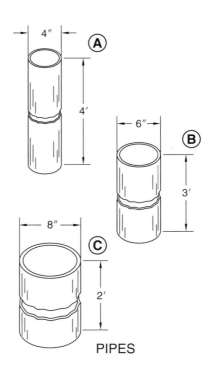

PIPES

12. A cylindrical tank has a diameter of 15″ and a length of 4′. A rectangular box measures 10″ × 15″ × 5′. Which has the largest capacity?

Name _____ Date _____

True-False and Completion

T   F   **1.** Solids have length, height, and depth.

T   F   **2.** The vertex is the common point of the triangular sides that form a pyramid.

T   F   **3.** A circle is a plane figure formed by a cutting plane parallel to the axis of a cone.

_____   **4.** _____ are solids bound by plane surfaces.

_____   **5.** A(n) _____ is a solid with two bases that are parallel and identical polygons.

_____   **6.** A(n) _____ is a solid generated by a straight line moving in contact with a curve and remaining parallel to the axis and its previous position.

_____   **7.** The _____ of a pyramid is the perpendicular distance from the vertex to the base.

_____   **8.** A(n) _____ is a solid generated by a circle revolving about one of its axes.

_____   **9.** _____ is the three-dimensional size of an object measured in cubic units.

T   F   **10.** A pyramid has a triangular base and rectangular sides.

T   F   **11.** A cubic foot contains 1728 cu in.

T   F   **12.** Cones have circular bases.

T   F   **13.** The sides of a prism are parallelograms.

T   F   **14.** The axis of a right cylinder is parallel to the base.

T   F   **15.** The circumference of a sphere is equal to the circumference of a great circle.

_____ **16.** A(n) _____ is a regular solid of four triangles.

_____ **17.** A(n) _____ is a prism with bases that are parallelograms.

_____ **18.** A hexahedron is a regular solid of six _____.

_____ **19.** A dodecahedron is a regular solid of _____ pentagons.

_____ **20.** The altitude of a prism is the perpendicular distance between two _____.

T   F   **21.** An oblique prism has lateral faces perpendicular to the bases.

T   F   **22.** A pyramid is named according to the kind of polygon forming the base.

T   F   **23.** Cones have circular bases.

T   F   **24.** An ellipse can be formed by passing a cutting plane through a cone.

_____ **25.** A(n) _____ is the remaining portion of a pyramid or cone with a cutting plane passed parallel to the base.

_____ **26.** A(n) _____ inch measures 1″ × 1″ × 1″ or its equivalent.

_____ **27.** A(n) _____ circle is formed by passing a cutting plane through the center of a sphere.

_____ **28.** A right pentagonal prism has _____ lateral faces.

T   F   **29.** Slant height is measured parallel to a side of a cone or pyramid.

T   F   **30.** A hyperbola is a plane figure formed by a cutting plane oblique to the axis and parallel to the elements of a cone.

These formulas may be used to solve the following problems:

Volume of a Rectangular Solid...............$V = l \times w \times h$

Volume of a Cylinder (Diameter)..$V = .7854 \times D^2 \times h$

Volume of a Cylinder (Radius)...................$V = \pi r^2 \times l$

Volume of a Sphere (Diameter).......................$V = \dfrac{\pi D^3}{6}$

Volume of a Sphere (Radius) ....$V = \dfrac{4\pi r^3}{3}$

Volume of a Cone:

   1. *Solve for area of base*.............. $A_b = .7854 \times D^2$

   2. *Solve for volume*.....................$V = \dfrac{A_b a}{3}$

Calculations

_____ **1.** A rectangular solid measures 8″ × 12″ × 72″. How many cubic feet does it contain?

_____ **2.** A hexahedron has a base length of 4.5″. How many cubic inches are in the hexahedron?

_____ **3.** A cylinder is 5 cm in length and 11 cm in diameter. What is the volume of the cylinder?

_____ **4.** A sphere has a diameter of 3.5″. What is its volume?

_____ **5.** A sphere has a radius of 1.25″. What is its volume?

6. Three cylinders measure as follows:
   Cylinder A—21″D × 32″l
   Cylinder B—20″D × 34″l
   Cylinder C—18″D × 40″l
   Which cylinder holds the largest volume of water?

7. Which holds the lesser volume, a 6″ × 6″ × 20″ rectangular solid or a 4″ radius × 20″ cylinder?

8. Worker A dug a 36″ diameter well, and Worker B dug a 30″ diameter well. Worker A struck water at 12′. Worker B struck water at 16′. Which worker removed the most dirt?

9. Which has the larger volume, a 3.25″ radius sphere or a 6.00″ diameter sphere?

10. The base of a cone has a 16″ diameter. The altitude of the cone is 28″. What is the volume of the cone? *Round to the next higher full cubic inch.*

11. An A-size carton measures 12″ × 12″ × 18″. What is the volume of the carton?

12. What is the volume of a 16″ diameter sphere?

Name _____ Date _____

These formulas may be used to solve the following problems:

Volume of a Rectangular Solid...............$V = l \times w \times h$

Volume of a Cylinder (Diameter)..$V = .7854 \times D^2 \times h$

Volume of a Cylinder (Radius)...................$V = \pi r^2 \times l$

Volume of a Sphere (Diameter).......................$V = \dfrac{\pi D^3}{6}$

Volume of a Sphere (Radius) ....$V = \dfrac{4\pi r^3}{3}$

Volume of a Cone:

  1. *Solve for area of base*............... $A_b = .7854 \times D^2$

  2. *Solve for volume*........................$V = \dfrac{A_b a}{3}$

_____ **1.** A storage room measures 46'-0" × 22'-0" with a 10'-0" ceiling. How many cubic yards does the storage room contain? *Round to the next higher full cubic yard.*

_____ **2.** A closet measures 3'-0" × 6'-0" × 8'-0". What is the storage capacity of the closet?

_____ **3.** What is the volume of a 2.25" radius sphere?

_____ **4.** Carton A is $12'' \times 16'' \times 24''$.
Carton B is $14'' \times 14'' \times 24''$.
Which has the larger capacity?

_____ **5.** A section of pipeline pipe
has an inside diameter of
$36''$. The pipe is $39'\text{-}6''$ long.
What is the capacity of the
pipe in cubic feet?

_____ **6.** A cone has a $20''$ diameter
base and a $32''$ altitude.
What is the volume of the
cone? _Round to the next
higher cubic foot._

_____ **7.** What is the volume of four
$3'$ radius spheres? _Round to
the nearest full cubic foot._

_____ **8.** A double-wide driveway
measures $20'\text{-}0''$ wide by $32'\text{-}0''$ long. The local building
code requires $4''$ of concrete.
How many full cubic yards
of concrete are required?

_____ **9.** A small dump truck has a $6' \times 8'$ bed with $30''$ sides.
How many cubic feet does
the dump truck hold?

_____ **10.** What is the volume of an
8 cm diameter sphere?

_____ **11.** A rectangular solid measures
8 cm $\times$ 8 cm $\times$ 16 cm. What
is the volume?

_____ **12.** A tank is $4'\text{-}0''$ in diameter
and $12'\text{-}0''$ long. What is the
volume?

A graph is a diagram that shows the relationship between two or more variables. Graphs present information in simple form. They are used to illustrate data and specifications. The most common types of graphs are the line graph, bar graph, circle graph, and pictograph.

LINE GRAPHS

A *graph* is a diagram that shows the relationship between two or more variables. A *line graph* is a graph in which points representing variables are connected by a line. Line graphs show the relationship and change between variables. Line graphs are also known as curves. See Figure 12-1.

Graph paper is used to plot line graphs. *Graph paper*, or coordinate paper, is paper with a number of intersecting vertical and horizontal lines. The lines on the paper are equal distances apart. Spaces on the coordinate paper are assigned values (number of units) according to the values of the variables of each problem.

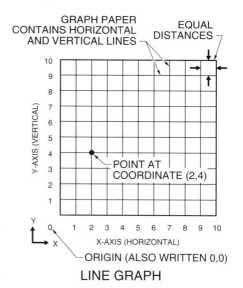

Figure 12-1. A line graph is a graph in which points representing variables are connected by a line.

An *axis* is a line serving as a basis from which to count the required number of spaces. The *x-axis* is the horizontal axis. The *y-axis* is the vertical axis. On a line graph there is one horizontal axis and one vertical axis.

The *origin* is the point where the two axes meet. It is labeled zero (0) or (0,0).

The first number represents the location on the x-axis. The second number represents the location on the y-axis.

The location of a point on a line graph is written as coordinates in parentheses. Coordinates give the x-axis to y-axis (x,y) relationship. For example, the point (2,4) is located two units from the origin on the x-axis and four units from the origin on the y-axis.

The three types of line graphs are straight-line, curved-line, and broken-line. See Figure 12-2.

The type of line on a line graph is dependent upon the relationship between the variables. Variables can have a causal relationship or no causal relationship.

Variables with a causal relationship are related to each other by a mathematical formula or rule. Variables with a causal relationship produce straight-line or curved-line graphs. An unknown value can be found if the value of the other variable is known.

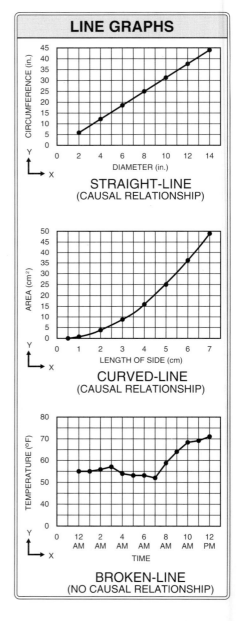

LINE GRAPHS

STRAIGHT-LINE
(CAUSAL RELATIONSHIP)

CURVED-LINE
(CAUSAL RELATIONSHIP)

BROKEN-LINE
(NO CAUSAL RELATIONSHIP)

Figure 12-2. The three types of line graphs are straight-line, curved-line, and broken-line.

For example, if the diameters of circles are 2″, 4″, 6″, 8″, etc., the circumferences are 6.28″, 12.56″, 18.84″, 25.12″, etc. When plotted with the x-axis as the diameter and the y-axis as the circumference, the points (2,6.28), (4,12.56), (6,18.84), and (8,25.12) produce a straight line.

If the numbers 1 through 7 are plotted on the x-axis and their squares are plotted on the y-axis, a curved-line graph is produced. The numbers on the y-axis are 1, 4, 9, 16, 25, 36, and 49.

Variables with no causal relationship are related to each other only through the line graph. Variables with no causal relationship produce broken-line graphs.

For example, a time-temperature graph shows temperature at a certain time. The temperature increased from approximately 52°F at 6:00 AM to approximately 71°F at 12:00 PM.

There is no set relationship, formula, or rule between the two variables of the time-temperature graph. To conserve space, the lower portion of the graph is not drawn.

Creating Line Graphs

To create a line graph, arrange the given values in a logical order. Organize the values from largest to smallest, smallest to largest, or the beginning of a time period to the end of a time period. Place the values in a table of values to organize them and simplify plotting. See Figure 12-3.

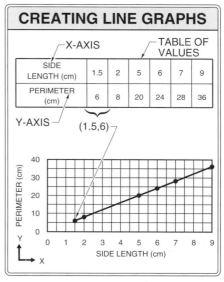

Figure 12-3. To create a line graph, place the values in a table of values.

For example, to plot the perimeter of squares ($P = 4s$) with side lengths of 7 cm, 9 cm, 2 cm, 6 cm, 1.5 cm, and 5 cm, multiply each side length by 4 to get perimeters of 28, 36, 8, 24, 6, and 20. Arrange the side lengths from smallest to largest in a table.

Decide which variable to locate on the x-axis and which to locate on the y-axis. The x-axis scale is on the bottom and the y-axis scale is to the left of the graph. Label the x-axis and y-axis and assign values to the spaces on the scales. The assigned values on the graph should conveniently represent the values to be plotted.

Plot each pair of values, or co-ordinates, by counting on the x-axis the value of *x* and by counting on the y-axis the value of *y*. Project the value of *x* horizontally and the value of *y* vertically. Place a point where the projections intersect. Once a number of points are plotted, connect the points with a straightedge or curve.

PRACTICE PROBLEMS

Line Graphs

1. Use the Current-Voltage Graph to find the current in the circuit if the voltage is 180 V.

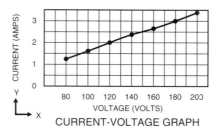

CURRENT-VOLTAGE GRAPH

2. Graph the double-pulley force-load information.

Force (x-axis) 4, 10, 11, 15, 16, 17.5, 20.5

Load (y-axis) 8, 20, 22, 30, 32, 35, 41

Note: Force equals load divided by 2 ($F = {}^{W}/_{2}$).

See Metro Population Growth Graph and answer questions 3–7.

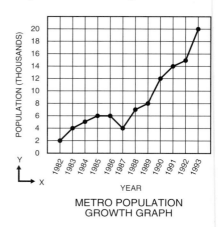

METRO POPULATION
GROWTH GRAPH

3. In which year was the growth rate flat?
4. In which year did the population grow by 4000?
5. What was the net population growth from 1983-1990?
6. Which year had the largest population increase?
7. In which year was there a decrease in population?

See Monthly Rainfall Graph and answer questions 8–10.

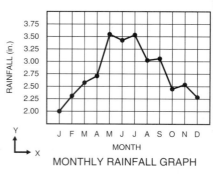

MONTHLY RAINFALL GRAPH

8. Which month had the least rainfall?
9. Which three-month period had the most rainfall?
10. The rainfall during January was _____ ″.

BAR GRAPHS

A *bar graph* is a graph in which values of variables are represented by bars. The bars may be drawn horizontally or vertically. The length of a bar indicates a value. See Figure 12-4.

To interpret a bar graph, determine the value of each space on the axis (scale). If the bars are horizontal, determine the space value on the horizontal scale. If the bars are vertical, determine the space value on the vertical scale. If the end of a bar is not on a graph line, the value is approximated.

Group and multiple bar graphs are used to represent data with more than one value for a variable. See Figure 12-5.

A *group bar graph* is a bar graph where each bar is divided into more defined variables. Each bar is divided into different sections which are either shaded, colored, or cross-hatched. Each section represents a value of the new variables.

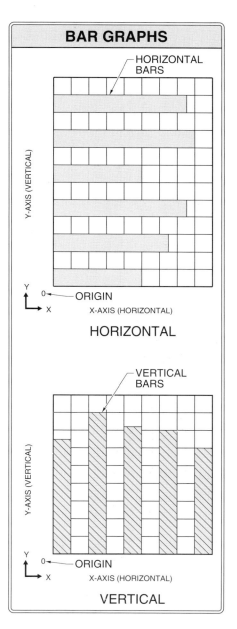

Figure 12-4. A bar graph is a graph in which values of variables are represented by bars.

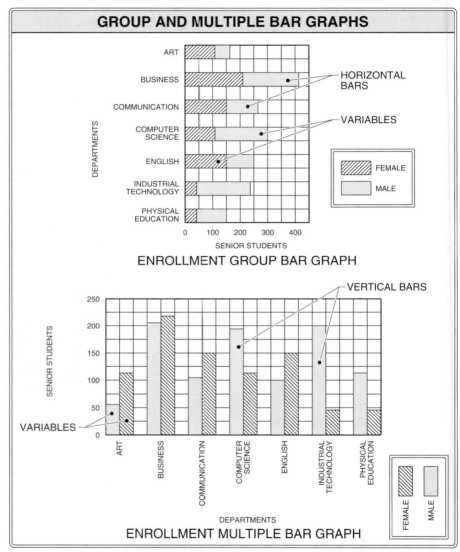

Figure 12-5. Group and multiple bar graphs are used to represent data with more than one value for a variable.

On the Enrollment Group Bar Graph, the total number of senior students enrolled in each department is represented by the whole bar (solid and cross-hatched section). The total enrollment is divided into the number of males (solid section) and the number of females (cross-hatched section). For example, the total number of students in the Industrial Technology Department is 240, 40 of whom are female and 200 are male.

A *multiple bar graph* is a bar graph where more than one bar is used to illustrate more defined variables. Each new bar is either shaded, colored, or cross-hatched. Each bar represents a value of the new variables.

On the Enrollment Multiple Bar Graph, the total number of senior students enrolled in each department is the sum of the bars of each variable. The total enrollment is divided into two bars which represent the number of males (solid section) and the number of females (cross-hatched section).

For example, the total number of senior students in the business department is the sum of the two bars, or 220 (male students) plus 210 (female students). The total number of students in the business department is 430.

Creating Bar Graphs

To create a bar graph, arrange the given values and variables in a logical order. Place the values in a table of values to organize them and simplify plotting. See Figure 12-6.

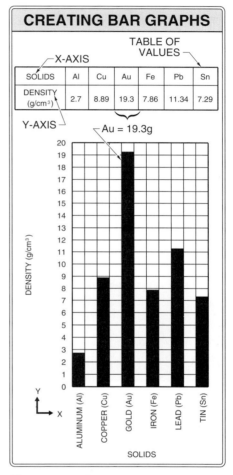

Figure 12-6. To create a bar graph, place the values in a table of values.

Decide which variable(s) to locate on the x-axis and which to locate on the y-axis. Label the x-axis and y-axis and assign the values to the spaces on the scales.

The assigned values on the graph should conveniently represent the values to be plotted. Draw each bar to the length or height cor-responding to its value.

For example, to create a bar graph of the density of solids where aluminum (Al) is 2.7 g/cm^3, iron (Fe) is 7.86 g/cm^3, lead (Pb) is 11.34 g/cm^3, copper (Cu) is 8.89 g/cm^3, gold (Au) is 19.3 g/cm^3, and tin (Sn) is 7.29 g/cm^3, arrange the solids in alphabetical order in a table. Label the x-axis as *Solids* and the y-axis as *Density*. Draw each bar to the length corresponding to its value.

PRACTICE PROBLEMS

Bar Graphs

See Motor Failure* Bar Graph and answer questions 1–5.

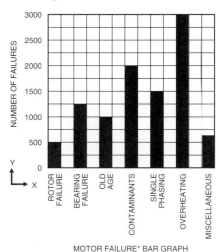

MOTOR FAILURE* BAR GRAPH

* 10,000 MOTOR
  FAILURES

1. More motors failed because of _____ than for any other reason.

2. _____ was the least-occur-ring cause of motor failure.

3. Contaminants and Bearing Fail-ure accounted for _____ motor failures.

4. _____ was the third largest cause of motor failure.

5. Single Phasing and _____ were the causes for the same number of motor failures.

6. Create a bar graph entitled Automobile Production Data Graph.

Year	Automobiles
1st	7,300,000
2nd	8,000,000
3rd	8,250,000
4th	9,800,000
5th	9,500,000
6th	8,400,000
7th	10,400,000

7. Create a horizontal group bar graph entitled Sales Data Graph.

Month	Salesperson	Units Sold
Jan	Jay	6
	Kris	3
	Kevin	4
Feb	Jay	2
	Kris	6
	Kevin	3
Mar	Jay	4
	Kris	6
	Kevin	4

See Technical Course Enrollments and answer questions 8–12.

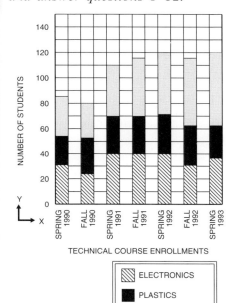

CIRCLE GRAPHS

A *circle graph* is a graph in which the circle represents 100% of a variable (total) and sectors of the circle represent parts of the total. Values are determined by the sizes of the sectors. Circle graphs are also known as pie charts. See Figure 12-7.

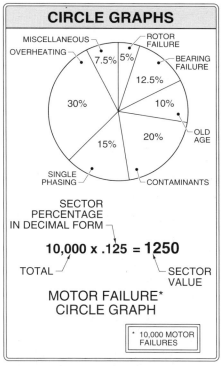

Figure 12-7. A circle graph is a graph in which the circle represents 100% of a variable and the sectors represent parts of the total.

8. The smallest enrollment area is _____.

9. Electronics enrollment increased by _____ from the fall of 1992 to the spring of 1993.

10. The total enrollment for the spring of 1991 was _____ students.

11. The largest growth in student enrollment from the spring of 1990 through the spring of 1993 occurred in the area of _____.

12. The term with the least total enrollment was _____.

A protractor is used to determine each angle of a sector. A circle contains 360°, which is 100% of the total. A 90° sector is equal to

$^{90}/_{360}$, which equals $^{1}/_{4}$, or 25%. The angle may also be found by multiplying it by .2$\overline{77}$.

For example, a 270° sector contains 75% ($270 \times .277 = 74.79 = 75\%$). All values of sectors are found by figuring what percentage of the whole the sector represents.

A circle graph usually has a percentage of a total in each sector. The percentage can be used to find the sector value if the total is known. To find the sector value when the total is known, multiply the total by the sector percentage in decimal form.

For example, to find the number of motor failures caused by bearing failure, multiply 10,000 by .125 ($10,000 \times .125 = 1250$). Of the 10,000 motor failures, 1250 were caused by bearing failure.

Creating Circle Graphs

To create a circle graph, place the values in table format. Divide each value by the total to get percentages in decimal form. Multiply 360° by each percentage to get the sector angle. Within a circle, lay out each angle with a protractor. See Figure 12-8.

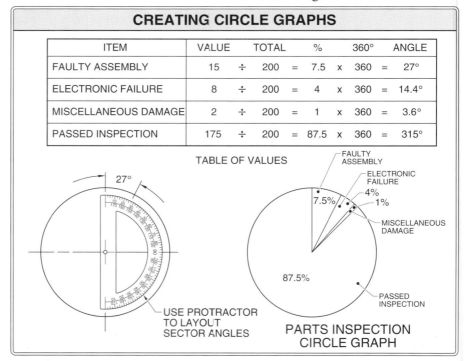

CREATING CIRCLE GRAPHS

ITEM	VALUE		TOTAL		%		360°		ANGLE
FAULTY ASSEMBLY	15	÷	200	=	7.5	x	360	=	27°
ELECTRONIC FAILURE	8	÷	200	=	4	x	360	=	14.4°
MISCELLANEOUS DAMAGE	2	÷	200	=	1	x	360	=	3.6°
PASSED INSPECTION	175	÷	200	=	87.5	x	360	=	315°

TABLE OF VALUES

27°

USE PROTRACTOR
TO LAYOUT
SECTOR ANGLES

FAULTY
ASSEMBLY

ELECTRONIC
FAILURE

4%

7.5%

1%

MISCELLANEOUS
DAMAGE

87.5%

PASSED
INSPECTION

PARTS INSPECTION
CIRCLE GRAPH

Figure 12-8. A table of values assists in creating a circle graph.

Lay out the smaller angles first. Each consecutive sector angle has a common side. The sum of all sector angles must equal 360°.

For example, in the Parts Inspection Circle Graph, 200 parts are inspected with 175 parts passing. Faulty Assembly, Electronic Failure, and Miscellaneous Damage account for the parts not passing inspection.

PRACTICE PROBLEMS

Circle Graphs

See Hardware Sales Circle Graph and answer questions 1–5.

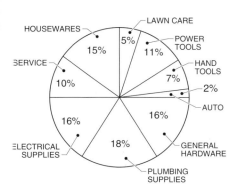

HOUSEWARES
LAWN CARE
5%
15%
POWER TOOLS
11%
SERVICE
HAND TOOLS
10%
7%
2%
AUTO
16%
16%
18%
ELECTRICAL SUPPLIES
GENERAL HARDWARE
PLUMBING SUPPLIES

HARDWARE SALES CIRCLE GRAPH

1. _____ represent the largest-selling product group.

2. Lawn Care products account for _____% of the total sales.

3. Electrical Supplies, Plumbing Supplies, and General Hardware account for _____% of the total sales.

4. _____ sales are smaller than any other product group.

5. General Hardware and _____ have equal percentages of sales.

Complete the PC-Tech Values Table, develop a circle graph entitled PC-Tech Manufacturing Co., and answer questions 6–12.

6. Marketing represents a(n) _____° sector on the circle graph.

7. Production represents a(n) _____° sector on the circle graph.

8. The Clerical workers comprise _____% of the total work force.

9. The second-largest department at PC-Tech Manufacturing Co. is _____.

10. Accounting represents a(n) _____° sector on the circle graph.

11. The total percentage of employees in the three smallest departments is _____%.

12. The total percentage of all workers, excluding Management, not in the Production department is _____%.

GROUP	VALUE		TOTAL	%		360°	ANGLE
MANAGEMENT	5	÷	=	x		=	
MARKETING	12	÷	=	x		=	
CLERICAL	7	÷	=	x		=	
ACCOUNTING	6	÷	=	x		=	
CUSTOMER SERVICE	3	÷	=	x		=	
MAINTENANCE	1	÷	=	x		=	
PRODUCTION	46	÷	=	x		=	

PC-TECH VALUES TABLE

PICTOGRAPHS

A *pictograph* is a graph in which pictures or symbols represent variables and their values. See Figure 12-9.

A pictograph shows approximations of values. In some cases the exact values are given. If exact values are not given and a partial symbol is used, the value is approximated. Whole symbols and partial symbols are used.

The symbol(s) have a visual relationship to the variable(s) on the graph. For example, oil barrels can be used for amounts of exported oil, cars can be used for automobile production, etc.

To interpret pictographs, the value of each symbol must be known. Count the number of symbols and partial symbols and multiply by the value of a symbol.

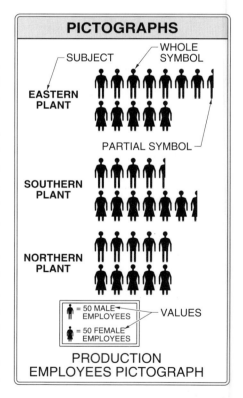

Figure 12-9. A pictograph is a graph in which pictures or symbols represent variables and their values.

In the Production Employees Pictograph, each symbol is equal to 50 employees. To find the number of employees at any plant listed, count the number of whole symbols, and multiply by 50. Add the approximate value of partial symbols.

For example, 625 people are employed at the Eastern Plant. The pictograph shows the relation in symbols (7½ male symbols × 50 = 375; 5 female symbols × 50 = 250; 375 + 250 = **625**).

Creating Pictographs

To create a pictograph, list the subject in the left column, and determine the symbols that will be used to represent variables. Assign a value to each symbol. Draw the number of whole and partial symbols required. See Figure 12-10.

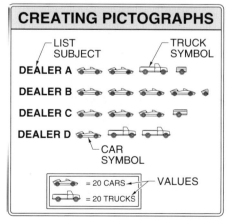

Figure 12-10. To create a pictograph, assign values to selected symbols.

For example, Dealer A sold 40 cars and 25 trucks, Dealer B sold approximately 85 cars, Dealer C sold 60 cars and 10 trucks, and Dealer D sold 20 cars and 40 trucks for the month. The number of vehicles sold is represented by the appropriate symbols.

PRACTICE PROBLEMS

Pictographs

See Annual Appliance Energy Usage* Pictograph and answer questions 1–6.

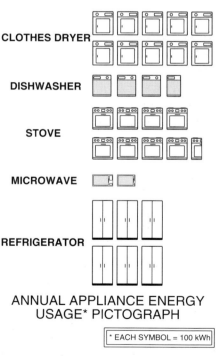

ANNUAL APPLIANCE ENERGY USAGE* PICTOGRAPH

* EACH SYMBOL = 100 kWh

1. The _____ have the lowest annual energy usage.

2. The _____ have the highest annual energy usage.
3. The annual energy usage of the Dishwashers is approximately _____ kWh.
4. The _____ have the second highest annual energy usage.
5. Refrigerators account for _____ kWh usage during the year.
6. Stoves consume _____ kWh during the year.

See Weekly Sandwich Sales* Pictograph and answer questions 7–12.

7. The restaurant sold _____ Hamburgers during the week.
8. BBQ Pork Sandwiches accounted for approximately _____ sandwiches sold during the week.
9. The restaurant sold approximately _____ Fish Sandwiches and Chicken Sandwiches during the week.

FISH SANDWICHES

HAMBURGERS

HOT DOGS

CHICKEN SANDWICHES

BBQ PORK SANDWICHES

WEEKLY SANDWICH SALES* PICTOGRAPH

* EACH SYMBOL = 100 SANDWICHES

10. _____ were the top seller during the week.
11. _____ more Hamburgers than Hot Dogs were sold during the week.
12. Total sandwich sales for the month were _____.

Name _____ Date _____

True-False and Completion

_____ **1.** A(n) _____ is a diagram that shows the relationship between two or more variables.

_____ **2.** The _____ is the point of a graph where the two axes meet.

T   F   **3.** Variables with a causal relationship are related to each other by a formula or rule.

T   F   **4.** The bars of a bar graph may be drawn vertically or horizontally.

_____ **5.** More than one bar is used to illustrate defined variables in a(n) _____ bar graph.

_____ **6.** A(n) _____ is used to determine each angle of a sector.

_____ **7.** A(n) _____ is a graph in which pictures or symbols represent variables and their values.

_____ **8.** Graph paper is also known as _____ paper.

T   F   **9.** The point (3,5) is located three units from the y-axis.

T   F   **10.** Variables with a causal relationship produce straight-line or curved-line graphs.

Calculations

See Insulation Step Voltage Test Graph and answer questions 1–5.

_____ **1.** Curve B shows a resistance of _____ MΩ (megohms) at 1 kV (kilovolt).

_____ **2.** Curve A shows a continual _____ in resistance.

_____ **3.** Curve A shows _____ MΩ when Curve B shows 10 MΩ.

_____ **4.** The maximum resistance shown on Curve B is _____ MΩ.

_____ **5.** The lowest reading on Curve A is approximately _____ MΩ.

See Public Opinion Poll Circle Graph and answer questions 6–10.

_____ **6.** The majority interviewed _____ with the question.

_____ **7.** No opinion is expressed by _____% of the people interviewed.

_____ **8.** Of those interviewed, _____% agree very strongly with the question.

_____ **9.** Of those interviewed, _____% disagree very strongly with the question.

_____ **10.** A total of _____% disagree or disagree very strongly with the question.

INSULATION STEP VOLTAGE TEST GRAPH

PUBLIC OPINION POLL CIRCLE GRAPH

Name _____ Date _____

Complete the Main Panelboard Table, develop a circle graph entitled
Main Panelboard Circuits Circle Graph, and answer questions 1–6.

ITEM	USE	VALUE	TOTAL	%	360°	ANGLE
15 A SPCB	GENERAL LIGHTING AND RECEPTACLES	6 ÷	=	x	=	
20 A SPCB	SPECIAL APPLIANCES	8 ÷	=	x	=	
20 A SPCB	SMALL APPLIANCES AND LAUNDRY	3 ÷	=	x	=	
20 A SPCB	SPARES	7 ÷	=	x	=	
30 A DPCB	OVENS AND COOKTOPS	2 ÷	=	x	=	
40 A DPCB	RANGES	1 ÷	=	x	=	
30 A DPCB	AC	1 ÷	=	x	=	
30 A DPCB	WATER HEATERS	1 ÷	=	x	=	
125 A DPCB	HEATING	1 ÷	=	x	=	
30 A DPCB	SPARES	2 ÷	=	x	=	

NOTE: SINGLE POLE CIRCUIT BREAKERS (SPCB) USED FOR 120 V CIRCUITS
DOUBLE POLE CIRCUIT BREAKERS (DPCB) USED FOR 240 V CIRCUITS

MAIN PANELBOARD TABLE

_____ **1.** Special Appliances constitute _____% of the CBs in the Main Panelboard.

_____ **2.** The Main Panelboard has a total of _____ SPCBs.

_____ **3.** Spares for 20 A SPCBs account for _____% of the total CBs. *(3 places)*

_____ **4.** A total of _____ CBs each account for angles of 11.25° of the circle graph.

_____ **5.** Ovens and Cooktops have the same number of circuits as _____.

_____ **6.** Ovens and Cooktops and _____ constitute the greatest use for DPCBs.

See Monthly Snowfall Graph and answer questions 7–10.

_____ **7.** More snow fell during _____ than during any other month.

_____ **8.** _____ and March had the least snowfall.

_____ **9.** A total of _____″ of snow fell from October through March.

_____ **10.** The snowfall during December was _____″.

See TV Age Groups Multiple Bar Graph and answer questions 11–20.

_____ **11.** The largest age group that watches Sports is _____ yrs.

_____ **12.** Children ages 6-9 yrs watch Sports and _____ equally.

_____ **13.** The least watched category is _____.

_____ **14.** The most watched category is _____.

_____ **15.** Situation Comedies are watched by _____ children ages 6-9.

_____ **16.** Movies are watched by _____ children ages 9-12.

_____ **17.** Sports are least watched by children _____ yrs old.

_____ **18.** _____ are watched by 200 children ages 9-12.

_____ **19.** Children ages _____ yrs watch the most cartoons.

_____ **20.** Situation Comedies are watched by _____ children ages 9-12.

MONTHLY SNOWFALL GRAPH

TC 6 YRS

6-9 YRS

9-12 YRS

TV AGE GROUPS
MULTIPLE BAR GRAPH

ANSWERS: Practice Problems

293

GRAPHS

Line Graphs 278
1. 3 A
3. 1985
5. 8000
7. 1986-1987
9. May, June, and July

Bar Graphs 282
1. Overheating
3. 3250
5. Bearing Failure
7. *See Sales Data Graph.*

9. 5
11. Computer Aided Manufacturing

Circle Graphs 285
1. Plumbing Supplies
3. 50
5. Electrical Supplies
7. 207
9. Marketing
11. 11.25

Pictographs 287
1. Microwaves
3. 375
5. 600
7. 1800
9. 650
11. 1100

APPENDIX

SYMBOLS

SYMBOL	TERM	SYMBOL	TERM
+	addition	′	minutes
∠	angle	×	multipli-cation
¢	cent	#	number
:	colon	‖	parallel
$\sqrt[3]{x}$	cube root	%	percent
°	degree	⊥	perpen-dicular
φ	diameter	π	pi (3.14)
÷,)‾, or)‾	division	±	plus or minus
$	dollar	$\overline{x}$	repetend
= or ∷	equal	∟	right angle
′	foot	″	seconds
>	greater than	$\sqrt{x}$	square root
″	inch	−	subtraction
<	less than	Δ	triangle

MULTIPLICATION TABLE

1	2	3	4	5	6	7	8	9	10	11	12
2	4	6	8	10	12	14	16	18	20	22	24
3	6	9	12	15	18	21	24	27	30	33	36
4	8	12	16	20	24	28	32	36	40	44	48
5	10	15	20	25	30	35	40	45	50	55	60
6	12	18	24	30	36	42	48	54	60	66	72
7	14	21	28	35	42	49	56	63	70	77	84
8	16	24	32	40	48	56	64	72	80	88	96
9	18	27	36	45	54	63	72	81	90	99	108
10	20	30	40	50	60	70	80	90	100	110	120

ROMAN NUMERALS

ARABIC	ROMAN	ARABIC	ROMAN	ARABIC	ROMAN
1	I	10	X	100	C
2	II	20	XX	200	CC
3	III	30	XXX	300	CCC
4	IV	40	XL	400	CD
5	V	50	L	500	D
6	VI	60	LX	600	DC
7	VII	70	LXX	700	DCC
8	VIII	80	LXXX	800	DCCC
9	IX	90	XC	900	DCCCC or CM
				1000	M
UNITS		TENS		HUNDREDS	

CIRCLES

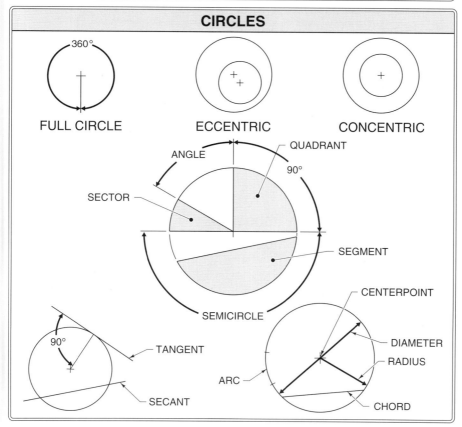

FULL CIRCLE   ECCENTRIC   CONCENTRIC

POLYGONS

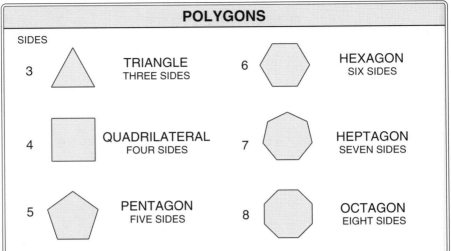

SIDES

3   **TRIANGLE**
THREE SIDES

6   **HEXAGON**
SIX SIDES

4   **QUADRILATERAL**
FOUR SIDES

7   **HEPTAGON**
SEVEN SIDES

5   **PENTAGON**
FIVE SIDES

8   **OCTAGON**
EIGHT SIDES

TRIANGLES

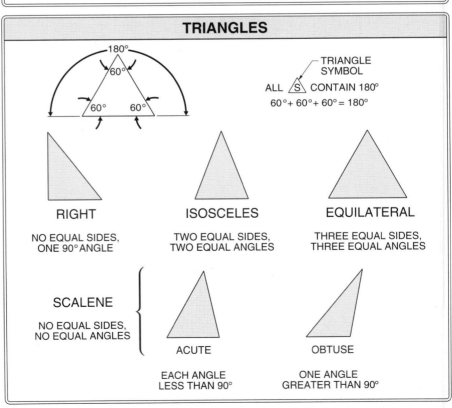

180°
60°
60° 60°

TRIANGLE
SYMBOL
ALL △S CONTAIN 180°
60° + 60° + 60° = 180°

RIGHT

NO EQUAL SIDES,
ONE 90° ANGLE

ISOSCELES

TWO EQUAL SIDES,
TWO EQUAL ANGLES

EQUILATERAL

THREE EQUAL SIDES,
THREE EQUAL ANGLES

SCALENE

NO EQUAL SIDES,
NO EQUAL ANGLES

ACUTE

EACH ANGLE
LESS THAN 90°

OBTUSE

ONE ANGLE
GREATER THAN 90°

PYTHAGOREAN THEOREM

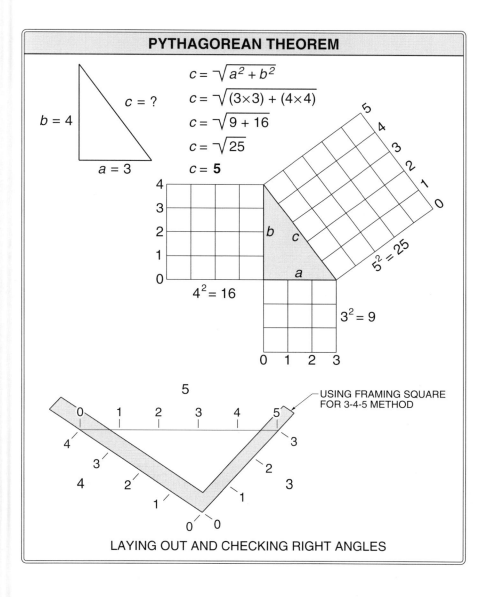

$c = \sqrt{a^2 + b^2}$

$c = \sqrt{(3 \times 3) + (4 \times 4)}$

$c = \sqrt{9 + 16}$

$c = \sqrt{25}$

$c = \mathbf{5}$

$b = 4$

$c = ?$

$a = 3$

$4^2 = 16$

$3^2 = 9$

$5^2 = 25$

USING FRAMING SQUARE FOR 3-4-5 METHOD

LAYING OUT AND CHECKING RIGHT ANGLES

QUADRILATERALS

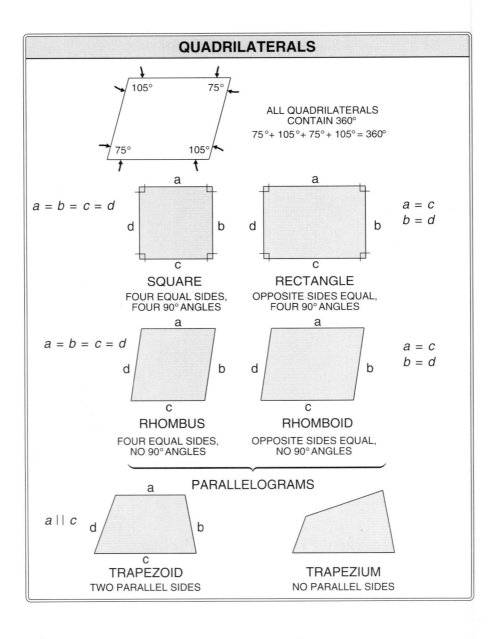

ALL QUADRILATERALS
CONTAIN 360°
75°+ 105°+ 75°+ 105° = 360°

$a = b = c = d$

SQUARE
FOUR EQUAL SIDES,
FOUR 90° ANGLES

$a = c$
$b = d$

RECTANGLE
OPPOSITE SIDES EQUAL,
FOUR 90° ANGLES

$a = b = c = d$

RHOMBUS
FOUR EQUAL SIDES,
NO 90° ANGLES

$a = c$
$b = d$

RHOMBOID
OPPOSITE SIDES EQUAL,
NO 90° ANGLES

PARALLELOGRAMS

$a \parallel c$

TRAPEZOID
TWO PARALLEL SIDES

TRAPEZIUM
NO PARALLEL SIDES

AREA

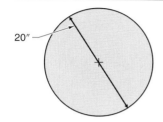

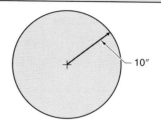

CIRCLE

CIRCUMFERENCE OF A CIRCLE (DIAMETER)
$C = \pi D$
$C = 3.1416 \times 20$
$C = $ **62.832″**

CIRCUMFERENCE OF A CIRCLE (RADIUS)
$C = 2\pi r$
$C = 2 \times 3.1416 \times 10$
$C = $ **62.832″**

AREA OF A CIRCLE (DIAMETER)
$A = .7854 \times D^2$
$A = .7854 \times (20 \times 20)$
$A = .7854 \times 400$
$A = $ **314.16 sq in.**

AREA OF A CIRCLE (RADIUS)
$A = \pi r^2$
$A = 3.1416 \times (10 \times 10)$
$A = 3.1416 \times 100$
$A = $ **314.16 sq in.**

AREA OF A SQUARE OR RECTANGLE
$A = l \times w$
$A = 16 \times 7$
$A = $ **112 sq in.**

AREA OF A TRIANGLE
$A = \frac{1}{2} bh$
$A = \frac{1}{2} \times (4 \times 5)$
$A = \frac{1}{2} \times 20$
$A = $ **10 sq in.**

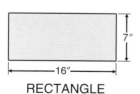

RECTANGLE

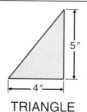

TRIANGLE

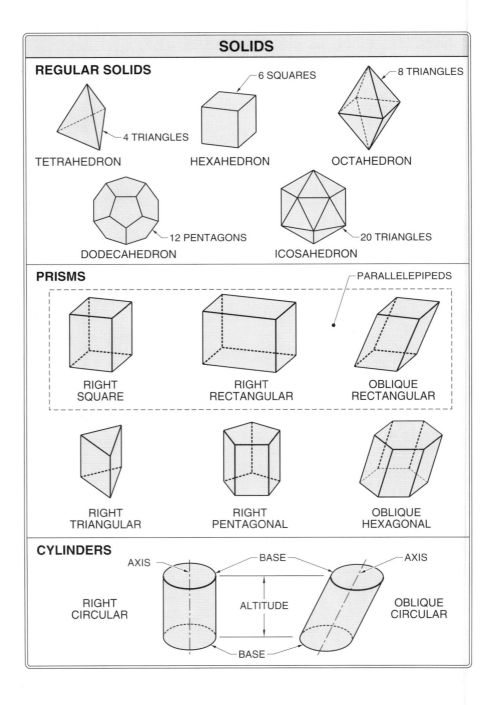

SOLIDS

REGULAR SOLIDS

6 SQUARES

8 TRIANGLES

4 TRIANGLES

TETRAHEDRON　　　HEXAHEDRON　　　OCTAHEDRON

12 PENTAGONS

20 TRIANGLES

DODECAHEDRON　　　ICOSAHEDRON

PRISMS

PARALLELEPIPEDS

RIGHT
SQUARE

RIGHT
RECTANGULAR

OBLIQUE
RECTANGULAR

RIGHT
TRIANGULAR

RIGHT
PENTAGONAL

OBLIQUE
HEXAGONAL

CYLINDERS

AXIS　　　BASE　　　AXIS

RIGHT
CIRCULAR

ALTITUDE

OBLIQUE
CIRCULAR

BASE

VOLUME . . .

VOLUME OF A RECTANGULAR SOLID
$V = l \times w \times h$
$V = 19 \times 10 \times 7$
$V = 1330$ sq in.
$V = \frac{1330}{144} = $ **9.236 sq ft**

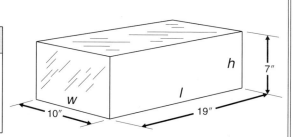

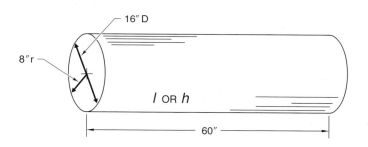

VOLUME OF A CYLINDER (DIAMETER)	VOLUME OF A CYLINDER (RADIUS)
$V = .7854 \times D^2 \times h$	$V = \pi r^2 \times l$
$V = .7854 \times (16 \times 16) \times 60$	$V = 3.1416 \times (8 \times 8) \times 60$
$V = .7854 \times 256 \times 60$	$V = 3.1416 \times 64 \times 60$
$V = 12{,}063.744$ cu in.	$V = 12{,}063.744$ cu in.
$V = \frac{12{,}063.744}{1728}$	$V = \frac{12{,}063.744}{1728}$
$V = $ **6.9813 cu ft**	$V = $ **6.9813 cu ft**

. . . VOLUME

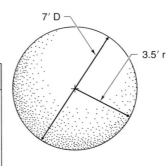

7' D

3.5' r

VOLUME OF A SPHERE (DIAMETER)

$$V = \frac{\pi D^3}{6}$$

$$V = \frac{3.1416 \times 7^3}{6}$$

$$V = \frac{3.1416 \times 343}{6}$$

$$V = \textbf{179.5948 cu ft}$$

VOLUME OF A SPHERE (RADIUS)

$$V = \frac{4\pi r^3}{3}$$

$$V = \frac{4 \times 3.1416 \times 3.5^3}{3}$$

$$V = \frac{4 \times 3.1416 \times 42.875}{3}$$

$$V = \textbf{179.5948 cu ft}$$

VOLUME OF A CONE

1. Solve for area of base

$$A_b = .7854 \times D^2$$

$$A_b = .7854 \times (3.25 \times 3.25)$$

$$A_b = .7854 \times 10.5625$$

$$A_b = \textbf{10.5625 sq ft}$$

2. Solve for volume

$$V = \frac{A_b a}{3}$$

$$V = \frac{10.5625 \times 5}{3}$$

$$V = \frac{52.8125}{3}$$

$$V = \textbf{17.6042 cu ft}$$

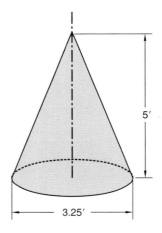

5'

3.25'

DECIMAL EQUIVALENTS

Fraction	Decimal	Fraction	Decimal
1/64	.015625	33/64	.515625
1/32	.03125	17/32	.53125
3/64	.046875	35/64	.546875
1/16	.0625	9/16	.5625
5/64	.078125	37/64	.578125
3/32	.09375	19/32	.59375
7/64	.109375	39/64	.609375
1/8	.125	5/8	.625
9/64	.140625	41/64	.640625
5/32	.15625	21/32	.65625
11/64	.171875	43/64	.671875
3/16	.1875	11/16	.6875
13/64	.203125	45/64	.703125
7/32	.21875	23/32	.71875
15/64	.234375	47/64	.734375
1/4	.250	3/4	.750
17/64	.265625	49/64	.765625
9/32	.28125	25/32	.78125
19/64	.296875	51/64	.796875
5/16	.3125	13/16	.8125
21/64	.328125	53/64	.828125
11/32	.34375	27/32	.84375
23/64	.359375	55/64	.859375
3/8	.375	7/8	.875
25/64	.390625	57/64	.890625
13/32	.40625	29/32	.90625
27/64	.421875	59/64	.921875
7/16	.4375	15/16	.9375
29/64	.453125	61/64	.953125
15/32	.46875	31/32	.96875
31/64	.484375	63/64	.984375
1/2	.500	1	1.000

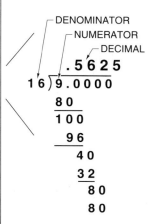

DENOMINATOR
NUMERATOR
DECIMAL

$$16 \overline{)9.0000} = .5625$$

```
      . 5 6 2 5
16 ) 9 . 0 0 0 0
     8 0
     1 0 0
       9 6
         4 0
         3 2
           8 0
           8 0
```

PERCENTAGE EQUIVALENTS

Percent	Proper Fraction	Decimal	Fractional Hundredth
5%	$\frac{1}{20}$	.05	$\frac{5}{100}$
10%	$\frac{1}{10}$	.10	$\frac{10}{100}$
12½%	$\frac{1}{8}$	.125	$\frac{12\frac{1}{2}}{100}$
15⅝%	$\frac{5}{32}$	.15375	$\frac{15\frac{5}{8}}{100}$
16⅔%	$\frac{1}{6}$	.16$\overline{66}$	$\frac{16\frac{2}{3}}{100}$
18¾%	$\frac{3}{16}$	.1875	$\frac{18\frac{3}{4}}{100}$
20%	$\frac{1}{5}$	.20	$\frac{20}{100}$
21⅞%	$\frac{7}{32}$	.21875	$\frac{21\frac{7}{8}}{100}$
25%	$\frac{1}{4}$	.25	$\frac{25}{100}$
28⅛%	$\frac{9}{32}$	.28125	$\frac{28\frac{1}{8}}{100}$
31¼%	$\frac{5}{16}$	.3125	$\frac{31\frac{1}{4}}{100}$
33⅓%	$\frac{1}{3}$	.33$\overline{33}$	$\frac{33\frac{1}{3}}{100}$
37½%	$\frac{3}{8}$	.375	$\frac{37\frac{1}{2}}{100}$
40%	$\frac{2}{5}$	.40	$\frac{40}{100}$
43¾%	$\frac{7}{16}$	.4375	$\frac{43\frac{3}{4}}{100}$
50%	$\frac{1}{2}$	.50	$\frac{50}{100}$
60%	$\frac{3}{5}$	.60	$\frac{60}{100}$
62½%	$\frac{5}{8}$	.625	$\frac{62\frac{1}{2}}{100}$
66⅔%	$\frac{2}{3}$	.66$\overline{66}$	$\frac{66\frac{2}{3}}{100}$
70%	$\frac{7}{10}$	.70	$\frac{70}{100}$
75%	$\frac{3}{4}$	.75	$\frac{75}{100}$
80%	$\frac{4}{5}$	.80	$\frac{80}{100}$
83⅓%	$\frac{5}{6}$	.83$\overline{33}$	$\frac{83\frac{1}{3}}{100}$
87½%	$\frac{7}{8}$	.875	$\frac{87\frac{1}{2}}{100}$
90%	$\frac{9}{10}$	.90	$\frac{90}{100}$

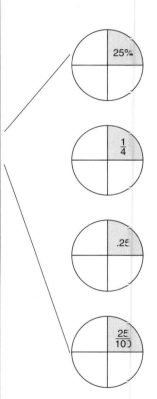

ENGLISH SYSTEM				
		Unit	**Abbr**	**Equivalents**
LENGTH		mile	mi	5280′, 320 rd, 1760 yd
		rod	rd	5.50 yd, 16.5′
		yard	yd	3′, 36″
		foot	ft *or* ′	12″, .333 yd
		inch	in. *or* ″	.083′, .028 yd
AREA		squaremile	sq mi *or* mi^2	640 A, 102,400 sq rd
		acre	A	4840 sq yd, 43,560 sq ft
		squarerod	sq rd *or* rd^2	30.25 sq yd, .00625 A
		squareyard	sq yd *or* yd^2	1296 sq in., 9 sq ft
		squarefoot	sq ft *or* ft^2	144 sq in., .111 sq yd
		squareinch	sq in. *or* in^2	.0069 sq ft, .00077 sq yd
VOLUME		cubicyard	cu yd *or* yd^3	27 cu ft, 46,656 cu in.
		cubicfoot	cu ft *or* ft^3	1728 cu in., .0370 cu yd
		cubicinch	cu in. *or* in^3	.00058 cu ft, .000021 cu yd
CAPACITY	*U.S. liquid measure*	gallon	gal.	4 qt (231 cu in.)
		quart	qt	2 pt (57.75 cu in.)
		pint	pt	4 gi (28.875 cu in.)
		gill	gi	4 fl oz (7.219 cu in.)
		fluidounce	fl oz	8 fl dr (1.805 cu in.)
		fluidram	fl dr	60 min (.226 cu in.)
		minim	min	⅙ fl dr (.003760 cu in.)
	U.S. dry measure	bushel	bu	4 pk (2150.42 cu in.)
		peck	pk	8 qt (537.605 cu in.)
		quart	qt	2 pt (67.201 cu in.)
		pint	pt	½ qt (33.600 cu in.)
	British imperial liquid and dry measure	bushel	bu	4 pk (2219.36 cu in.)
		peck	pk	2 gal. (554.84 cu in.)
		gallon	gal.	4 qt (277.420 cu in.)
		quart	qt	2 pt (69.355 cu in.)
		pint	pt	4 gi (34.678 cu in.)
		gill	gi	5 fl oz (8.669 cu in.)
		fluidounce	fl oz	8 fl dr (1.7339 cu in.)
		fluidram	fl dr	60 min (.216734 cu in.)
		minim	min	¹⁄₆₀ fl dr (.003612 cu in.)
MASS AND WEIGHT	*avoirdupois*	ton		2000 lb
		shortton	t	2000 lb
		longton		2240 lb
		pound	lb *or* #	16 oz, 7000 gr
		ounce	oz	16 dr, 437.5 gr
		dram	dr	27.344 gr, .0625 oz
		grain	gr	.037 dr, .002286 oz
	troy	pound	lb	12 oz, 240 dwt, 5760 gr
		ounce	oz	20 dwt, 480 gr
		pennyweight	dwt *or* pwt	24 gr, .05 oz
		grain	gr	.042 dwt, .002083 oz
	apothecaries'	pound	lb ap	12 oz, 5760 gr
		ounce	oz ap	8 dr ap, 480 gr
		dram	dr ap	3 s ap, 60 gr
		scruple	s ap	20 gr, .333 dr ap
		grain	gr	.05 s, .002083 oz, .0166 dr ap

ENGLISH TO METRIC EQUIVALENTS

		Unit	Metric Equivalent
LENGTH		mile	1.609 km
		rod	5.029 m
		yard	.9144 m
		foot	30.48 cm
		inch	2.54 cm
AREA		square mile	2.590 k^2
		acre	.405 hectacre, 4047 m^2
		square rod	25.293 m^2
		square yard	.836 m^2
		square foot	.093 m^2
		square inch	6.452 cm^2
VOLUME		cubic yard	.765 m^3
		cubic foot	.028 m^3
		cubic inch	16.387 cm^3
CAPACITY	*U.S. liquid measure*	gallon	3.785 l
		quart	.946 l
		pint	.473 l
		gill	118.294 ml
		fluidounce	29.573 ml
		fluidram	3.697 ml
		minim	.061610 ml
	U.S. dry measure	bushel	35.239 l
		peck	8.810 l
		quart	1.101 l
		pint	.551 l
	British imperial liquid and dry measure	bushel	.036 m^3
		peck	.0091 m^3
		gallon	4.546 l
		quart	1.136 l
		pint	568.26 cm^3
		gill	142.066 cm^3
		fluidounce	28.412 cm^3
		fluidram	3.5516 cm^3
		minim	.059194 cm^3
MASS AND WEIGHT	*avoidupois*	short ton	.907 t
		long ton	1.016 t
		pound	.454 kg
		ounce	28.350 g
		dram	1.772 g
		grain	.0648 g
	troy	pound	.373 kg
		ounce	31.103 g
		pennyweight	1.555 g
		grain	.0648 g
	apothecaries'	pound	.373 kg
		ounce	31.103 g
		dram	3.888 g
		scruple	1.296 g
		grain	.0648 g

METRIC SYSTEM			
	Unit	**Abbr**	**Number of Base Units**
LENGTH	kilometer	km	1000
	hectometer	hm	100
	dekameter	dam	10
	*meter	m	1
	decimeter	dm	.1
	centimeter	cm	.01
	millimeter	mm	.001
AREA	square kilometer	sq km *or* km^2	1,000,000
	hectare	ha	10,000
	are	a	100
	square centimeter	sq cm *or* cm^2	.0001
VOLUME	cubic centimeter	cu cm, cm^3, *or* cc	.000001
	cubic decimeter	dm^3	.001
	*cubic meter	m^3	1
CAPACITY	kiloliter	kl	1000
	hectoliter	hl	100
	dekaliter	dal	10
	*liter	l	1
	cubic decimeter	dm^3	1
	deciliter	dl	.10
	centiliter	cl	.01
	milliliter	ml	.001
MASS AND WEIGHT	metric ton	t	1,000,000
	kilogram	kg	1000
	hectogram	hg	100
	dekagram	dag	10
	*gram	g	1
	decigram	dg	.10
	centigram	cg	.01
	milligram	mg	.001

*Base Units

METRIC TO ENGLISH EQUIVALENTS

	Unit	British Equivalent		
LENGTH	kilometer	.62 mi		
	hectometer	109.36 yd		
	dekameter	32.81′		
	meter	39.37″		
	decimeter	3.94″		
	centimeter	.39″		
	millimeter	.039″		
AREA	square kilometer	.3861 sq mi		
	hectacre	2.47 A		
	are	119.60 sq yd		
	square centimeter	.155 sq in.		
VOLUME	cubic centimeter	.061 cu in.		
	cubic decimeter	61.023 cu in.		
	cubic meter	1.307 cu yd		
		cubic	*dry*	*liquid*
CAPACITY	kiloliter	1.31 cu yd		
	hectoliter	3.53 cu ft	2.84 bu	
	dekaliter	.35 cu ft	1.14 pk	2.64 gal.
	liter	61.02 cu in.	.908 qt	1.057 qt
	cubic decimeter	61.02 cu in.	.908 qt	1.057 qt
	deciliter	6.1 cu in.	.18 pt	.21 pt
	centiliter	.61 cu in.		338 fl oz
	milliliter	.061 cu in.		.27 fl dr
MASS AND WEIGHT	metric ton	1.102 t		
	kilogram	2.2046 lb		
	hectogram	3.527 oz		
	dekagram	.353 oz		
	gram	.035 oz		
	decigram	1.543 gr		
	centigram	.154 gr		
	milligram	.015 gr		

METRIC PREFIXES

Multiples and Submultiples	Prefixes	Symbols	Meaning
$1,000,000,000,000 = 10^{12}$	tera	T	trillion
$1,000,000,000 = 10^{9}$	giga	G	billion
$1,000,000 = 10^{6}$	mega	M	million
$1000 = 10^{3}$	kilo	k	thousand
$100 = 10^{2}$	hecto	h	hundred
$10 = 10^{1}$	deka	d	ten
Unit $1 = 10^{0}$			
$.1 = 10^{-1}$	deci	d	tenth
$.01 = 10^{-2}$	centi	c	hundredth
$.001 = 10^{-3}$	milli	m	thousandth
$.000001 = 10^{-6}$	micro	μ	millionth
$.000000001 = 10^{-9}$	nano	n	billionth

METRIC CONVERSIONS

Initial Units	Final Units											
	giga	mega	kilo	hecto	deka	base unit	deci	centi	milli	micro	nano	pico
giga		3R	6R	7R	8R	9R	10R	11R	12R	15R	18R	21R
mega	3L		3R	4R	5R	6R	7R	8R	9R	12R	15R	18R
kilo	6L	3L		1R	2R	3R	4R	5R	6R	9R	12R	15R
hecto	7L	4L	1L		1R	2R	3R	4R	5R	8R	11R	14R
deka	8L	5L	2L	1L		1R	2R	3R	4R	7R	10R	13R
base unit	9L	6L	3L	2L	1L		1R	2R	3R	6R	9R	12R
deci	10L	7L	4L	3L	2L	1L		1R	2R	5R	8R	11R
centi	11L	8L	5L	4L	3L	2L	1L		1R	4R	7R	10R
milli	12L	9L	6L	5L	4L	3L	2L	1L		3R	6R	9R
micro	15L	12L	9L	8L	7L	6L	5L	4L	3L		3R	6R
nano	18L	15L	12L	11L	10L	9L	8L	7L	6L	3L		3R
pico	21L	18L	15L	14L	13L	12L	11L	10L	9L	6L	3L	

R = Move the decimal point to the right.   L = Move the decimal point to the left.

POWERS, ROOTS, and RECIPROCALS...

No.	Square	Cube	Sq Root	Cu Root	Reciprocal	No.
1	1	1	1.00000	1.00000	1.0000000	1
2	4	8	1.41421	1.25992	.5000000	2
3	9	27	1.73205	1.44225	.3333333	3
4	16	64	2.00000	1.58740	.2500000	4
5	25	125	2.23607	1.70998	.2000000	5
6	36	216	2.44949	1.81712	.1666667	6
7	49	343	2.64575	1.91293	.1428571	7
8	64	512	2.82843	2.00000	.1250000	8
9	81	729	3.00000	2.08008	.1111111	9
10	100	1000	3.16228	2.15443	.1000000	10
11	121	1331	3.31662	2.22398	.0909091	11
12	144	1728	3.46410	2.28943	.0833333	12
13	169	2197	3.60555	2.35133	.0769231	13
14	196	2744	3.74166	2.41014	.0714286	14
15	225	3375	3.87298	2.46621	.0666667	15
16	256	4096	4.00000	2.51984	.0625000	16
17	289	4913	4.12311	2.57128	.0588235	17
18	324	5832	4.24264	2.62074	.0555556	18
19	361	6859	4.35890	2.66840	.0526316	19
20	400	8000	4.47214	2.71442	.0500000	20
21	441	9261	4.58258	2.75892	.0476190	21
22	484	10,648	4.69042	2.80204	.0454545	22
23	529	12,167	4.79583	2.84387	.0434783	23
24	576	13,824	4.89898	2.88450	.0416667	24
25	625	15,625	5.00000	2.92402	.0400000	25
26	676	17,576	5.09902	2.96250	.0384615	26
27	729	19,683	5.19615	3.00000	.0370370	27
28	784	21,952	5.29150	3.03659	.0357143	28
29	841	24,389	5.38516	3.07232	.0344828	29
30	900	27,000	5.47723	3.10723	.0333333	30
31	961	29,791	5.56776	3.14138	.0322581	31
32	1024	32,768	5.65685	3.17480	.0312500	32
33	1089	35,937	5.74456	3.20753	.0303030	33

No.	Square	Cube	Sq Root	Cu Root	Reciprocal	No.
	...POWERS, ROOTS, and RECIPROCALS...					
35	1225	42,875	5.91608	3.27107	.0285714	35
36	1296	46,656	6.00000	3.30193	.0277778	36
37	1369	50,653	6.08276	3.33222	.0270270	37
38	1444	54,872	6.16441	3.36198	.0263158	38
39	1521	59,319	6.24500	3.39121	.0256410	39
40	1600	64,000	6.32456	3.41995	.0250000	40
41	1681	68,921	6.40312	3.44822	.0243902	41
42	1764	74,088	6.48074	3.47603	.0238095	42
43	1849	79,507	6.55744	3.50340	.0232558	43
44	1936	85,184	6.63325	3.53035	.0227273	44
45	2025	91,125	6.70820	3.55689	.0222222	45
46	2116	97,336	6.78233	3.58305	.0217391	46
47	2209	103,823	6.85565	3.60883	.0212766	47
48	2304	110,592	6.92820	3.63424	.0208333	48
49	2401	117,649	7.00000	3.65931	.0204082	49
50	2500	125,000	7.07107	3.68403	.0200000	50
51	2601	132,651	7.14143	3.70843	.0196078	51
52	2704	140,608	7.21110	3.73251	.0192308	52
53	2809	148,877	7.28011	3.75629	.0188679	53
54	2916	157,464	7.34847	3.77976	.0185185	54
55	3025	166,375	7.41620	3.80295	.0181818	55
56	3136	175,616	7.48331	3.82586	.0178571	56
57	3249	185,193	7.54983	3.84850	.0175439	57
58	3364	195,112	7.61577	3.87088	.0172414	58
59	3481	205,379	7.68115	3.89300	.0169492	59
60	3600	216,000	7.74597	3.91487	.0166667	60
61	3721	226,981	7.81025	3.93650	.0163934	61
62	3844	238,328	7.87401	3.95789	.0161290	62
63	3969	250,047	7.93725	3.97906	.0158730	63
64	4096	262,144	8.00000	4.00000	.0156250	64
65	4225	274,625	8.06226	4.02073	.0153846	65
66	4356	287,496	8.12404	4.04124	.0151515	66
67	4489	300,763	8.18535	4.06155	.0149254	67
68	4624	314,432	8.24621	4.08166	.0147059	68

...POWERS, ROOTS, and RECIPROCALS

No.	Square	Cube	Sq Root	Cu Root	Reciprocal	No.
69	4761	328,509	8.30662	4.10157	.0144928	69
70	4900	343,000	8.36660	4.12129	.0142857	70
71	5041	357,911	8.42615	4.14082	.0140845	71
72	5184	373,248	8.48528	4.16017	.0138889	72
73	5329	389,017	8.54400	4.17934	.0136986	73
74	5476	405,224	8.60233	4.19834	.0135135	74
75	5625	421,875	8.66025	4.21716	.0133333	75
76	5776	438,976	8.71780	4.23582	.0131579	76
77	5929	456,533	8.77496	4.25432	.0129870	77
78	6084	474,552	8.83176	4.27266	.0128205	78
79	6241	493,039	8.88819	4.29084	.0126582	79
80	6400	512,000	8.94427	4.30887	.0125000	80
81	6561	531,441	9.00000	4.32675	.0123457	81
82	6724	551,368	9.05539	4.34448	.0121951	82
83	6889	571,787	9.11043	4.36207	.0120482	83
84	7056	592,704	9.16515	4.37952	.0119048	84
85	7225	614,125	9.21954	4.39683	.0117647	85
86	7396	636,056	9.27362	4.41400	.0116279	86
87	7569	658,503	9.32738	4.43105	.0114943	87
88	7744	681,472	9.38083	4.44797	.0113636	88
89	7921	704,969	9.43398	4.46475	.0112360	89
90	8100	729,000	9.48683	4.48140	.0111111	90
91	8281	753,571	9.53939	4.49794	.0109890	91
92	8464	778,688	9.59166	4.51436	.0108696	92
93	8649	804,357	9.64365	4.53065	.0107527	93
94	8836	830,584	9.69536	4.54684	.0106383	94
95	9025	857,375	9.74679	4.56290	.0105263	95
96	9216	884,736	9.79796	4.57886	.0104167	96
97	9409	912,673	9.84886	4.59470	.0103093	97
98	9604	941,192	9.89949	4.61044	.0102041	98
99	9801	970,299	9.94987	4.62607	.0101010	99
100	10,000	1,000,000	10.00000	4.64159	.0100000	100

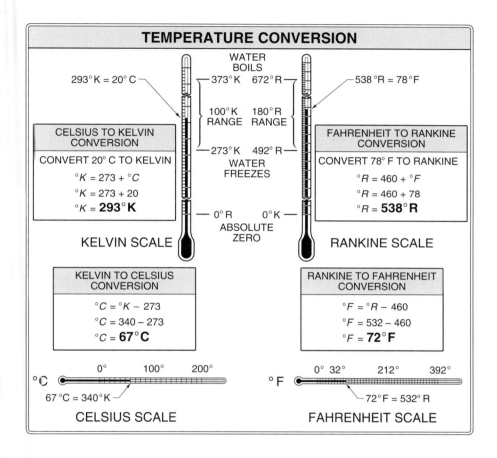

TEMPERATURE CONVERSION

WATER BOILS

293° K = 20° C ── ─373° K 672° R─ ──538 °R = 78 °F

100° K 180° R
RANGE RANGE

CELSIUS TO KELVIN CONVERSION

─273° K 492° R─

WATER FREEZES

FAHRENHEIT TO RANKINE CONVERSION

CONVERT 20° C TO KELVIN
$°K = 273 + °C$
$°K = 273 + 20$
$°K = \mathbf{293°K}$

CONVERT 78° F TO RANKINE
$°R = 460 + °F$
$°R = 460 + 78$
$°R = \mathbf{538°R}$

── 0° R   0° K ──
ABSOLUTE ZERO

KELVIN SCALE

RANKINE SCALE

KELVIN TO CELSIUS CONVERSION

RANKINE TO FAHRENHEIT CONVERSION

$°C = °K - 273$
$°C = 340 - 273$
$°C = \mathbf{67°C}$

$°F = °R - 460$
$°F = 532 - 460$
$°F = \mathbf{72°F}$

0°   100°   200°

°C

67 °C = 340° K ─

CELSIUS SCALE

0° 32°   212°   392°

°F

─72° F = 532° R

FAHRENHEIT SCALE

GLOSSARY

Terms in this glossary are defined as they relate to mathematics.

A

absolute zero: A theoretical condition where no heat is present.

abstract number: Any number used by itself.

acute angle: An angle that contains less than 90°. See *angle*.

acute angle

acute triangle: A scalene triangle with each angle less than 90°. See *scalene triangle*.

adding: The process of uniting two or more numbers to make one number.

adjacent angles: Angles that have the same vertex and one side in common. See *vertex*.

altitude: 1. The perpendicular dimension from the vertex to the base of a triangle, pyramid, or cone. **2.** The perpendicular distance between the two bases of a prism. See *vertex* and *base*.

angle: The intersection of two lines or sides.

apothecaries' fluid measure: The measurement of drugs.

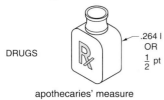

apothecaries' measure

Arabic numerals: Numerals expressed by the ten digits 0, 1, 2, 3, 4, 5, 6, 7, 8, and 9.

arc: A portion of the circumference of a circle. See *circumference*.

area: The number of units of length multiplied by the number of units of width. Area is expressed in square units.

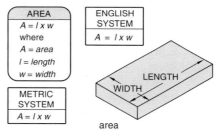

area

axis: 1. In solid figures, an imaginary line about which a geometric figure rotates. **2.** In graphs, the horizontal (x) or vertical (y) coordinates. **3.** A line serving as a basis from which to count the required number of spaces on a graph.

avoirdupois weight: The measurement of heavy weights. Used for objects such as coal, grain, livestock, etc.

avoirdupois weight

B

bar graph: A graph in which values of variables are represented by bars.

base: 1. In mathematics, a repeated factor that is multiplied to obtain a power. **2.** In plane figures, the side upon which a triangle stands. **3.** In solid figures, the ends of a prism. See *factor* and *power*.

C

cancellation: The process of dividing one factor in the dividend and one factor in the divisor by the same number. See *dividend* and *divisor.*

capacity: The measure of volume that a container can hold.

causes: Actions that produce effects.

centerpoint: The point a circle or arc is drawn around. See *circle* and *arc.*

chord: In a circle, a line from circumference to circumference not through the centerpoint. See *circumference.*

circle: 1. In plane figures, a figure generated about a centerpoint. **2.** In solid figures, a figure formed by a cutting plane perpendicular to the axis of a cone.

circle graph: A graph in which the circle represents 100% of a variable (total), and sectors of the circle represent parts of the total.

circumference: The boundary of a circle.

commission: A percentage of the total sale (total money received for a business transaction) which is paid to the individual who performs the transaction.

commission rate: The percentage used to figure commissions. See *commission.*

common denominator: The denominator of a group of fractions that has the same denominator. See *denominator.*

common factors: Two or more ordinary factors that divide into two or more numbers an exact number of times. See *factor.*

complementary angles: Two angles

complementary angles

formed by three lines in which the sum of the two angles equals 90°. See *angle.*

complex fraction: A fraction that has a fraction, improper fraction, mixed number, or mathematical process in its numerator, denominator, or both. See *improper fraction, mixed number, numerator,* and *denominator.*

compound denominate number: A denominate number that has more than one unit of measure. See *denominate number.*

compound proportion: A proportion where some of the terms are products of two variables.

compound ratio: Two ratio expressions in which the first terms are multiplied, the second terms are multiplied, and the relation of the first product to the second product is stated. See *ratio* and *product.*

concentric circles: Two or more circles with different diameters but the same centerpoint. See *circle, diameter,* and *centerpoint.*

concrete number: Any number used with a name of something or a unit of measure.

cone: A solid generated by a straight line moving in contact with a curve and passing through the vertex. See *vertex.*

cone

conic section: A curve produced by a plane intersecting a right circular cone. See *right circular cone.*

continuous quantities: Numbers resulting from measurements that cannot be directly counted but are close approximations.

cube root: A number which has three equal roots or factors. See *root* and *factor*.

cubic inch: A unit of measure that equals $1'' \times 1'' \times 1''$.

cubic foot: A unit of measure that equals $12'' \times 12'' \times 12'' = 1728$ cu in.

cubic measure: The measurement of volume used to find the three-dimensional size of an object. See *volume*.

cubic yard: A unit of measure that equals $3' \times 3' \times 3' = 27$ cu in.

curved line: A line that continually changes direction.

cylinder: A solid generated by a straight line (genatrix) moving in contact with a curve and remaining parallel to the axis and its previous position. See *axis* and *parallel*.

cylinder

D

decimal: A number expressed in base 10.

decimal fraction: A fraction with the denominator 10, 100, 1000, etc. See *denominator*.

decimal point: The period at the left of a proper decimal number or the period that separates the parts of a mixed decimal. See *proper decimal number* and *mixed decimal number*.

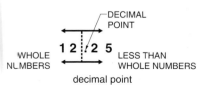

decimal point

denominate number: A concrete number that has one unit of measure. See *concrete number*.

denominator: The lower portion of a fraction which shows into how many parts the whole number is divided.

$$\frac{3}{4} \leftarrow$$

denominator

diameter: The distance from circumference to circumference through the centerpoint of a circle. See *circumference* and *centerpoint*.

difference: The result of subtraction. See *subtraction*.

direct proportion: A statement of equality between two ratios in which the first of four terms divided by the second equals the third divided by the fourth. See *ratio*.

direct ratio: A simple ratio that results when the first term is divided by the second term. See *ratio*.

discount: The reduction (amount of money) of the marked price.

dividend: The number to be divided.

dividing: The process of finding how many times one number contains another number.

divisor: The number that the dividend is divided by. See *dividend*.

dodecahedron: A regular solid of twelve pentagons.

dry measure: The measurement of dry items, such as grain, vegetables, etc.

dry measure

E

eccentric circles: Two or more circles with different diameters and different center-points. See *circle, diameter*, and *center-point*.

effects: The results of actions or causes.

ellipse: A plane figure formed by a cutting plane oblique to the axis of a cone, but at a greater angle with the axis than with the elements of the cone. See *plane figure* and *axis*.

equation: A means of showing that two numbers or two groups of numbers are equal to the same amount.

equilateral triangle: A triangle that has three equal angles and three equal sides. Each angle is 60°. See *angle*.

equilateral triangle

even number: Any number that can be divided by 2 an exact number of times.

┌─ CAN BE
│ DIVIDED BY 2

2, 4, 6, 8, 10

even number

evolution: The general process of finding roots of numbers. See *root*.

exponent: Indicates the number of times the base is raised to a power (number of times used as a factor). See *base* and *power*.

extension: The property of a body by which it occupies space and has one or more of the three dimensions of length, width, and thickness.

extracting (the square root): The process of finding one of the two equal factors of a number. See *factor*.

extremes: The two outer numbers of a proportion. See *proportion*.

F

factor: A number being multiplied.

formula: A mathematical equation that contains a fact, rule, or principle. See *equation*.

fraction: One part of a whole number. See *whole number*.

fractional hundredth: A fraction having a denominator of 100 and a numerator equal to the percent. See *fraction, denominator* and *numerator*.

frustum: The remaining portion of a pyramid or cone with a cutting plane passed parallel to the base. See *cone* and *pyramid*.

G

graph: A diagram that shows the relationship between two or more variables.

graph paper: Paper with a number of intersecting vertical and horizontal lines.

great circle: The circle formed by passing a cutting plane through the center of a sphere. See *circle* and *sphere*.

group bar graph: A bar graph where each bar is divided into more defined variables.

H

heptagon: A seven-sided plane figure. See *plane figure*.

SEVEN SIDES

heptagon

hexagon: A six-sided plane figure. See *plane figure*.

SIX SIDES

hexagon

hexahedron: A regular solid of six squares. See *regular solid.*

highest common factor: The largest common factor. See *factor.*

horizontal: Parallel to the horizon.

horizontal line: A line that is parallel to the horizon.

hyperbola: A plane figure formed by a cutting plane that has a smaller angle with the axis than with the elements of a cone. See *plane figure, axis,* and *cone.*

hypotenuse: The side of a right triangle opposite the right angle. See *right triangle.*

I

icosahedron: A regular solid of twenty triangles. See *regular solid.*

improper fraction: A fraction with the numerator larger than the denominator. See *numerator* and *denominator.*

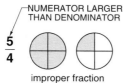

NUMERATOR LARGER THAN DENOMINATOR

$\frac{5}{4}$

improper fraction

inclined line: A line that is slanted and is neither horizontal nor vertical. See *horizontal* and *vertical.*

index: The small figure to the left of the root sign.

integer: A number that has no fractional or decimal part.

ALSO KNOWN AS WHOLE NUMBER

1, 2, 3, 4, 5

integer

interest rate: The percent of the principal paid back. See *principal.*

inverse proportion: A proportion in which an increase in one quantity results in a proportional decrease in the other related quantity. See *proportion.*

inverse ratio: A simple ratio that results when the second term is divided by the first term. See *ratio.*

involution: The process of finding powers of numbers by raising the base to the indicated power (exponent). See *base* and *power.*

irregular polygon: A polygon with unequal sides and unequal angles. See *polygon.*

irregular polyhedra: Solids with faces that are irregular polygons (unequal sides). See *polygon.*

isosceles triangle: A triangle that contains two equal angles and two equal sides. See *triangle.*

isosceles triangle

J

K

L

line: The boundary of a surface.

line graph: A graph in which points representing variables are connected by a line.

linear measure: The measurement of length used to find the one-dimensional length of an object.

liquid measure: The measurement of liquids, such as water or fuel.

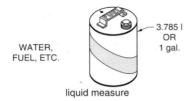

WATER,
FUEL, ETC.

3.785 l
OR
1 gal.

liquid measure

lowest common denominator (LCD): The smallest number into which the denominators of a group of two or more fractions divides an exact number of times. See *denominator* and *fraction*.

M

marked price: The price at which products are originally intended to be sold.

mass: The measurement of matter contained in an object.

means: The two inner numbers of a proportion. See *proportion*.

minuend: The number that is subtracted from. See *subtraction*.

mixed decimal number: A decimal number that has a whole number and a decimal number separated by a decimal point. See *whole number*.

CAN BE DIVIDED
BY ITSELF AND 1

1, 2, 3, 5, 7

prime number

mixed number: A combination of a whole number and a proper fraction. See *whole number* and *proper fraction*.

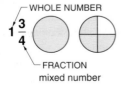

WHOLE NUMBER

$1\frac{3}{4}$

FRACTION

mixed number

multiple bar graph: A bar graph where more than one bar is used to illustrate more defined variables.

multiple discount: The discount found by applying more than one percent discount to the marked price, resulting in the net price.

multiplicand: The number being multiplied. See *multiplying*.

multiplier: The number multiplied by. See *multiplying*.

multiplying: The process of adding one number as many times as there are units in the other.

N

net price: The amount remaining after discount is subtracted from marked price. See *marked price*.

net proceeds: The amount of the total sale left after the commission is deducted. See *commission*.

numerator: Upper portion of a fraction which shows the number of parts taken from the denominator. See *denominator*.

$$\frac{3}{4}$$

numerator

O

oblique cylinder: A cylinder with the axis not perpendicular to the base. See *cylinder* and *perpendicular*.

obtuse angle: An angle that contains more than 90°. See *angle*.

MORE
THAN 90°

obtuse angle

obtuse triangle: A scalene triangle with one angle greater than 90°. See *scalene triangle* and *angle*.

octagon: An eight-sided plane figure. See *plane figure*.

EIGHT SIDES

octagon

octahedron: A regular solid of eight triangles. See *regular solid*.

odd number: Any number that cannot be divided by 2 an exact number of times.

CANNOT BE
DIVIDED BY 2

1, 3, 5, 7, 9

odd number

ordinary number: A number that is not a prime number. See *prime number*.

ordinary factor: Any number whether it is odd, even, or prime. See *odd number, even number,* and *prime number*.

origin: The point where the two axes meet. See *axis*.

P

parabola: A plane figure formed by a cutting plane oblique to the axis and parallel to the elements of a cone. See *plane figure, axis, parallel,* and *cone*.

parallel: Remaining the same distance apart.

parallelepiped: A prism with bases that are parallelograms. See *prism* and *parallelogram*.

parallel lines: Two or more lines that remain the same distance apart.

parallelogram: A four-sided plane figure with opposite sides parallel and equal. See *plane figure* and *parallel*.

pentagon: A five-sided plane figure. See *plane figure*.

FIVE SIDES

pentagon

percent: One part of 100 parts.

percentage: A method of expressing a part of a whole number in terms of hundredths. See *whole numbers*.

perfect square: A number that has a whole number square root. See *whole number* and *square root*.

period: A group of three places that is separated from the other periods by a comma.

perpendicular: At a 90° angle.

perpendicular line: A line that makes a 90° angle with another line.

pictograph: A graph in which pictures, or symbols, represent variables and their values.

place: The position that a digit occupies and that represents the value of the digit.

plane figure: A flat figure with no depth.

plumb: An exact verticality (determined by a plumb bob and line) with the Earth's surface. See *vertical*.

polygon: A many-sided plane figure. See *plane figure*.

polyhedra: Solids bound by plane surfaces (faces).

power: The product of a repeated factor. See *product* and *factor*.

prime factor: A factor consisting of a prime number. See *prime number*.

prime number: Number that can be divided an exact number of times only by itself and the number 1. See *factor* and *prime number.*

CAN BE DIVIDED
BY ITSELF AND 1

1, 2, 3, 5, 7

prime number

principal: The amount of money borrowed.

prism: A solid with two bases that are parallel and identical polygons. See *base, parallel,* and *polygon.*

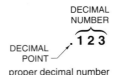

prism

product: The result of multiplication. See *multiplying.*

proper decimal number: A decimal number that has no whole numbers. See *whole number.*

DECIMAL
NUMBER

. 1 2 3

DECIMAL
POINT

proper decimal number

proper fraction: A fraction with the denominator larger than the numerator. See *denominator* and *numerator.*

$$\frac{5}{8}$$

DENOMINATOR LARGER
THAN NUMERATOR

proper fraction

proportion: An expression of equality between two ratios. See *ratio.*

pyramid: A solid with a base that is a polygon and sides that are triangles. See *base, polygon,* and *triangle.*

pyramid

Pythagorean Theorem: A theorem stating that the square of the hypotenuse of a right triangle is equal to the sum of the squares of the other two sides. See *hypotenuse* and *right triangle.*

Q

quadrilateral: A four-sided polygon with four interior angles that total 360°. See *polygon.*

quadrant: One-fourth of a circle containing 90°.

quotient: The result of division. See *dividing.*

R

radical sign: The symbol used to indicate the root of a number. See *root.*

radicand: The number under the radical sign from which the root is to be found. See *radical sign* and *root.*

radius: The distance from the centerpoint to the circumference of a circle. See *centerpoint* and *circumference.*

ratio: The relationship between two quantities or terms.

ratio transformation: The change of the ratio form. See *ratio.*

rectangle: A quadrilateral with opposite sides equal and four 90° angles.

rectangular parallelepiped: A prism with bases and faces that are all rectangular. See *prism.*

regular polygon: A polygon with equal sides and equal angles. See *polygon.*

regular polyhedra: A solid with faces that are regular polygons (equal sides). See *polygon.*

regular pyramid: A pyramid with a base that is a regular polygon and vertex that is perpendicular to the base. See *pyramid* and *vertex.*

regular solid: A solid with faces that are regular polygons (equal sides). See *polygons.*

relation: How much larger or smaller one term is in comparison with another term.

remainder: The part of the dividend left over when the quotient is not a whole number. See *dividend* and *quotient.*

repeating decimal: A decimal with a repetend. See *repetend.*

repetend: A group of figures of a decimal number that are repeated infinitely.

rhomboid: A quadrilateral with opposite sides equal and no 90° angles. See *quadrilateral.*

rhombus: A quadrilateral with all sides equal and no 90° angles. See *quadrilateral.*

right angle: Two lines that intersect perpendicular to each other. See *perpendicular.*

right angle

right circular cone: A cone with the axis at a 90° angle to the circular base. See *cone, axis,* and *base.*

right cylinder: A cylinder with the axis perpendicular to the base. See *cylinder, axis,* and *base.*

right parallelepiped: A prism with all edges perpendicular to the bases. See *parallelepiped, prism,* and *base.*

right prism: A prism with lateral faces perpendicular to the bases. See *prism* and *base.*

right triangle: A triangle that contains one 90° angle and no equal sides.

right triangle

Roman numerals: Numerals expressed by the letters I, V, X, L, C, D, and M.

root (of a number): One of the equal factors of a power. See *factor* and *power.*

S

scalene triangle: A triangle that has no equal angles or equal sides. See *triangle.*

scalene triangles

secant: A straight line touching the circumference of a circle at two points. See *circumference.*

sector: A pie-shaped piece of a circle.

segment: The portion of a circle set off by a chord. See *chord.*

semicircle: One-half of a circle containing 180°.

similar fractions: Fractions that have a common denominator. See *common denominator.*

simple interest: The amount of money charged for money borrowed (loan).

simple ratio: The ratio between two terms. See *ratio*.

slant height: The distance from the base to the vertex parallel to a side (of a pyramid). See *base* and *vertex*.

small circle: The circle formed by passing a cutting plane through a sphere but not through the center. See *circle* and *sphere*.

sphere: A solid generated by a circle revolving about one of its axes. See *circle* and *axis*.

sphere

square: A quadrilateral with all sides equal and four 90° angles. See *quadrilateral*.

square inch: An area that measures 1″ × 1″ or its equivalent.

square foot: Contains 144 sq in. (12″ × 12″ = 144 sq in.).

square measure: The measurement of area used to find the two-dimensional size of plane (flat) surfaces.

square root: Two equal factors or roots of a number. See *factor* and *root*.

straight angle: Two lines that intersect to form a straight line.

straight angle

straight line: The shortest distance between two points.

subtracting: The process of taking one number away from another number (the opposite of adding).

subtrahend: The number to be subtracted. See *subtracting*.

sum: The result of addition. See *adding*.

superscript: A small figure above and to the right of the base. See *base*.

supplementary angles: Two angles formed by three lines in which the sum of the two angles equals 180°. See *angle*.

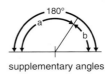

supplementary angles

T

tangent: A straight line touching the curve of the circumference of a circle at only one point. See *circumference* and *circle*.

tax: A charge paid on incomes, products, and services as a means of raising money to pay the expenses of city, county, state, and federal governments.

tax rate: The percentage of the selling prices, value of property, etc.

temperature: The measurement of hotness or coldness.

terms: The numerator and denominator of a fraction. See *numerator* and *denominator*.

tetrahedron: A regular solid of four triangles. See *regular solid*.

thermometer: An instrument used for measuring temperature.

time: The amount of time allotted for repayment of the principal plus interest.

total sale: The marked price plus the tax. See *marked price* and *tax*.

trapezium: A quadrilateral with no sides parallel. See *quadrilateral.*

trapezoid: A quadrilateral with two sides parallel. See *quadrilateral.*

triangle: Three-sided polygon with three interior angles. See *polygon.*

THREE SIDES

triangle

troy weight: The measurement of precious metals, such as gold, silver, etc.

GOLD, SILVER, ETC.

troy weights

U

unit of measure: 1. The specific item(s) being added, subtracted, multiplied, or divided. **2.** A standard by which a quantity such as length, area, capacity, or weight is measured. See *area, capacity,* and *weight.*

V

variable: A quantity that may change.

vertex: 1. The point of intersection of the sides of an angle. **2.** The common point of the triangular sides that forms a pyramid. See *pyramid.*

vertical: Perpendicular to the horizon. See *perpendicular.*

vertical line: A line that is perpendicular to the horizon. See *perpendicular.*

volume: 1. The contents of a three-dimensional object. **2.** The three-dimensional size of an object measured in cubic units.

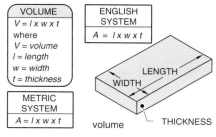

VOLUME
$V = l \times w \times t$
where
$V = volume$
$l = length$
$w = width$
$t = thickness$

METRIC SYSTEM
$A = l \times w \times t$

ENGLISH SYSTEM
$A = l \times w \times t$

LENGTH

WIDTH

volume

THICKNESS

W

weight: A measure of gravity or the force of the Earth's attraction for a body.

whole number: Any number that has no fractional or decimal part.

ALSO KNOWN AS INTEGER

1, 2, 3, 4, 5

whole number

X

x-axis: The horizontal axis.

Y

y-axis: The vertical axis.

Z

INDEX